# Exercise Programming for Cardiac Patients

Jerzy Rybicki · Adam Staron
Editors

# Exercise Programming for Cardiac Patients

Springer

*Editors*
Jerzy Rybicki
Repty Rehabilitation Center
Tarnowskie Góry, Poland

Adam Staron
The Upper Silesian Medical Center
Katowice, Poland

ISBN 978-3-031-88474-0 ISBN 978-3-031-88475-7 (eBook)
https://doi.org/10.1007/978-3-031-88475-7



This Springer imprint is published by the registered company Springer Nature Switzerland AG
The registered company address is: Gewerbestrasse 11, 6330 Cham, Switzerland

*We appreciate our families' patience and understanding of the significant time we have dedicated to completing this task, without whom this book would not have been possible.*

# Preface

Cardiac rehabilitation as a multi-component and comprehensive intervention reduces mortality and hospital readmissions and improves physical capacity and quality of life. It has the highest class of recommendation as an efficacious treatment of patients with myocardial infarction, after coronary revascularization, with chronic coronary syndrome, in patients with heart failure, and as cardiovascular disease prevention in clinical practice. Exercise training and physical activity counseling remain critical components of contemporary comprehensive multistage cardiac rehabilitation programs, including clinical evaluation, management of cardiovascular risk factors, dietary education, and psychological, social, and vocational support. Both duration and program settings (residential, outpatient, and hybrid) can differ between countries. However, exercise training encompasses aerobic, strength, flexibility, and neuromotor components. Our objective is to offer a reader a practical approach to exercise programming for patients with cardiovascular diseases. This position has been enriched with numerous tables and figures to minimize the text and facilitate understanding of the content. We hope it will serve as a critical resource for exercise practitioners, offering practical guidance on implementing exercise programming for a broad spectrum of patient groups. We firmly believe that the art of exercise prescription is the best cardiovascular therapy.

Katowice, Poland
Tarnowskie Góry, Poland

Adam Staron
Jerzy Rybicki

# Acknowledgments

We want to express our deepest gratitude to the publishers for their support throughout this journey. We thank our families for their patience and encouragement. We want to thank the reviewers for their valuable insight into the content and comments on improving this book's quality. We are in outstanding debt to the contributing authors for sharing their expertise. We wish to thank Mr. Rayno Klassen for participating in the photographs demonstrating the exercise training in practice.

Lastly, I would like to thank Mentor and Teacher, Dr. Jerzy Rybicki, for his extraordinary commitment to completing this project.

Adam Staron

# Contents

# Editors and Contributors

## About the Editors

**Jerzy Rybicki** MD, Ph.D., is an Internal medicine and rehabilitation specialist. He worked at Repty Rehabilitation Center in Tarnowskie Gory, Poland, between 1977 and 2017, where he served as Head of the 50-bed Cardiac Rehabilitation Department between 2006 and 2017. As the Board Member of the Cardiac Rehabilitation and Exercise Physiology Group of the Polish Cardiac Society, he has been actively involved in the development of cardiac rehabilitation programs. Dr. Rybicki is an author and co-author of 6 textbooks published in Polish language and 63 peer-reviewed articles. He developed individual ramp tests on a treadmill that attain a specific number of METs exactly at 10 min of the test. Dr. Rybicki invented exercise intensity calculators facilitating exercise programming. He has been Co-Director of an international Cardiac Rehabilitation Workshop held since 2023.

**Adam Staron** MD, Ph.D., Cardiologist with over 10 years' experience in cardiac rehabilitation. In 2014, he implemented as first in Poland yoga and tai chi in the Cardiac Rehabilitation Department in Katowice. Dr. Staron established the Cardiac Rehabilitation Unit at Prince Sultan Cardiac Center in Riyadh, Saudi Arabia, and he has written as co-author the curriculum of the Cardiac Rehabilitation Fellowship for certified cardiologists and physiatrists, approved by the Saudi Council for Health Specialties (2020).

In 2022, as a main author, he published a textbook: "Exercise Prescription in Cardiac Rehabilitation," published by MDPI. Dr. Staron has been Co-Director of the International Cardiac Rehabilitation Workshop held since 2023. Currently, he is working as the Head of the Cardiac Rehabilitation Department at the Upper Silesian Medical Center, Katowice, Poland. Dr. Staron serves as a Board Member of the Cardiac Rehabilitation and Exercise Physiology Group of the Polish Cardiac Society.

In the past, Dr. Staron's area of interest was novel techniques of echocardiography; he was awarded the European Society of Echocardiography training grant in 2008 (3-dimensional and speckle-tracking echocardiography).

## Contributors

**Amal Albatini** Ministry of Health Kuwait, Sulaibkhat, Kuwait

**Adré Rossouw** Clinical Exercise Physiologist Cardia Clinic, Jabriya, Kuwait

**Jerzy Rybicki** "REPTY" Upper Silesian Rehabilitation Center Tarnowskie Góry Poland, Silesian Center for Heart Disease, Zabrze, Poland

**Adam Staron** The Upper Silesian Medical Center, Katowice, Poland

# Abbreviations

| | |
|---|---|
| 1-RM | One-repetition maximum |
| 2-MST | Two-minute step test |
| 6MWD | Six-minute walk distance |
| 6MWSpeed | Six-minute walk speed |
| 6MWT | Six-minute walk test |
| AACVPR | American Association of Cardiovascular and Pulmonary Rehabilitation |
| ABI | Ankle brachial index |
| ACS | Acute coronary syndromes |
| ACSM | American College of Sports Medicine |
| ADL | Activities of daily living |
| AF | Atrial fibrillation |
| AHA | American Heart Association |
| AR | Active recovery |
| AT | An anaerobic threshold |
| AV | Atrio-ventricular |
| BF | Breathing frequency |
| BiVAD | Biventricular assist device |
| BM | Body mass |
| BMI | Body mass index |
| BP | Blood pressure |
| BR | Breathing reserve |
| CABG | Coronary artery bypass grafting |
| CACR | Canadian Association of Cardiac Rehabilitation |
| CAD | Coronary artery disease |
| CHF | Chronic heart failure |
| CP | Critical power |
| CPET | Cardiopulmonary exercise testing |
| CPET-ESE | Cardiopulmonary exercise testing and echocardiography |
| CR | Cardiac rehabilitation |
| CRF | Cardiorespiratory fitness |

| | |
|---|---|
| CRT | Cardiac resynchronization therapy |
| CRT-D | Cardiac resynchronization therapy with defibrillator |
| CVD | Cardiovascular disease |
| DBP | Diastolic blood pressure |
| DM | Diabetes mellitus |
| EACPR | European Association of Cardiovascular Prevention and Rehabilitation |
| ECG | Electrocardiogram |
| EE | Energy expenditure |
| ESC | European Society of Cardiology |
| EXPERT | Exercise Prescription in Everyday Practice & Rehabilitative Training |
| ExT | Exercise test |
| FEV1 | Forced expiratory volume in 1 second |
| FITT | Frequency, intensity, time, type |
| FITT-VP | Frequency, intensity, time, type, volume, and progression |
| FRIEND | Fitness Registry and the Importance of Exercise National Database |
| G | Grade |
| HDL | High-density lipoproteins |
| HF | Heart failure |
| HFmrEF | Heart failure with mildly reduced ejection fraction |
| HFpEF | Heart failure with preserved ejection fraction |
| HFrEF | Heart failure with reduced ejection fraction |
| HIIT | High-intensity interval training |
| HR | Heart rate |
| HRR | Heart rate reserve |
| HT | Heart transplantation |
| ICD | Implantable cardioverter-defibrillator |
| ICED | Implantable cardiac electrical devices |
| IMT | Inspiratory muscle training |
| ISWT | Incremental shuttle walking test |
| ITC | Interval-to-continuous |
| LDL | Low-density lipoproteins |
| l-HIIT | Long high-intensity interval training |
| LICT | Low-intensity continuous training |
| LIIT | Low-intensity interval training |
| L-M | Low-to-moderate |
| LP | Last positive |
| LV | Left ventricle |
| LVAD | Left ventricular assist device |
| LVEF | Left ventricular ejection fraction |
| MAP | Mean arterial pressure |
| MET | Metabolic equivalent of task |
| MET-R | Reserve of metabolic equivalents of task |

| | |
|---|---|
| MGA | Multidimensional geriatric assessment |
| m-HIIT | Moderate high-intensity interval training |
| MI | Myocardial infarction |
| MICT | Moderate-intensity continuous training |
| MIIT | Moderate-intensity interval training |
| MIT | Myocardial ischemic threshold |
| $mL \times kg^{-1} \times min^{-1}$ | Milliliters per kilogram per minute |
| MSEC | Maximal short-term exercise capacity |
| M-V | Moderate-to-vigorous |
| MVV | Maximal voluntary ventilation |
| NMES | Neuromuscular electrical stimulation |
| NSTEMI | Non-ST elevation myocardial infarction |
| NYHA | New York Heart Association |
| OUES | Oxygen uptake efficiency slope |
| PA | Physical activity |
| PAD | Peripheral arterial disease |
| PCI | Percutaneous coronary intervention |
| PEF | Peak expiratory flow |
| $P_{ET}CO_2$ | End-tidal carbon dioxide partial pressure |
| $P_{ET}O_2$ | End-tidal oxygen partial pressure |
| PPM | Permanent cardiac pacemaker |
| PR | Passive recovery |
| PRET | Perceptually regulated exercise training |
| PWR | Peak work rate |
| RER | Respiratory exchange ratio |
| ROM | Range of motion |
| RPE | Rating of perceived exertion |
| RT | Resistance training |
| S | Speed |
| SBP | Systolic blood pressure |
| s-HIIT | Short high-intensity interval training |
| SPPB | Short Physical Performance Battery |
| STEMI | ST Elevation myocardial infarction |
| TAVI | Transcatheter aortic valve implantation |
| TG | Triglycerides |
| THR | Training heart rate |
| TMET | Training metabolic equivalent of task |
| TT | Talk test |
| Tv | Tidal volume |
| VAD | Ventricular assist device |
| VC | Vital capacity |
| Vd | Dead space |
| VE | Minute ventilation |
| $VO_2$ | Oxygen uptake |
| VT | Ventilatory threshold |

| | |
|---|---|
| W | Watts |
| WR | Work rate |

# List of Figures

# List of Tables

# Chapter 1
# Functional Capacity Assessment

**Jerzy Rybicki and Adam Staron**

**Abstract** Functional capacity assessment is critical to evaluate risk profile and individual patient needs in the cardiac rehabilitation (CR) program. The goal is to ensure safe exercise training is provided with minimal risk entry. Such an evaluation will also support goal setting for the patient. Before the exercise training, a field test, a symptom-limited electrocardiographic or cardiopulmonary exercise test, is undertaken for prognostic, diagnostic, and therapeutic purposes, representing minimal, routine, and optimal testing standards, respectively. However, the need to perform a cardiopulmonary test is crucial to providing the data for meticulous exercise intensity prescription. In practice, the types of testing conducted by exercise professionals vary, depending on financial issues, local routine, and individual approaches. At the end of the supervised exercise training program, another evaluation should be performed to check individual progress, verify the goals and prescribe the long-term program.

Functional assessment can be performed as follows:

- Field test
  - six-minute walk test (6MWT)
  - incremental shuttle walk test (ISWT)
  - two-min step test (2MST).
- Electrocardiographic exercise test (ExT)
- Cardio-pulmonary exercise test (CPET).

Cardiopulmonary exercise tests are recognized as the "gold standard" for direct measurement of cardiorespiratory fitness (CRF); however, due to their limited availability, electrocardiographic exercise testing remains the most common diagnostic tool in exercise laboratories. In such tests, however, the dedicated formulas estimate

J. Rybicki
"REPTY" Upper Silesian Rehabilitation Center Tarnowskie Góry Poland, Silesian Center for Heart Disease, Zabrze, Poland
e-mail: j_rybicki@cyberia.pl

A. Staron (✉)
The Upper Silesian Medical Center, Katowice, Poland
e-mail: astaron1973@wp.pl



J. Rybicki and A. Staron (eds.), *Exercise Programming for Cardiac Patients*,
https://doi.org/10.1007/978-3-031-88475-7_1

maximal oxygen uptake indirectly. Field tests should be performed as complementary to conventional testing, offering just approximate functional capacity assessment.

**Keywords** Functional capacity · Metabolic equivalent of task · Cardiopulmonary exercise testing · Ventilatory threshold · Field tests

## 1.1 Electrocardiographic Exercise Test

Depending on the exercise method, exercise tests are usually conducted on a treadmill or leg cycle ergometer. Treadmills represent a physiological way of movement; however, they require more space and are more expensive than cycle ergometers. Electrically or mechanically braked cycle ergometers allow for a better ECG signal than treadmills; however, the tests are typically terminated due to the fatigue of the quadriceps muscles. In addition, the maximal oxygen uptake attained on the leg ergometer is lower by 10–22% compared with the treadmill [1, 2].

The standard physical performance measure is the maximum or peak minute oxygen uptake (VO2max or VO2peak, respectively) during gradually increasing exercise. The method of measuring VO2 divides exercise tests into ExT, in which VO2 expressed in metabolic equivalents of task (MET) is estimated using regression equations based on peak mechanical parameters (speed and incline of a treadmill or bicycle load), and CPET, in which direct measurement of VO2 is performed.

The ACSM formula for walking and running is applied to estimate maximal VO2 during the treadmill tests [3]:

$$\text{Walking: } VO_2 = (0.1 \times S) + (1.8 \times S \times G) + 3.5 \tag{1.1}$$

$$\text{Running: } VO_2 = (0.2 \times S) + (0.9 \times S \times G) + 3.5 \tag{1.2}$$

where: $VO_2$, oxygen uptake expressed in mL × $kg^{-1}$ × $min^{-1}$; S, speed in m/min; G, grade expressed in decimal format (e.g. 10% = 0.10); 3.5, resting component in mL × $kg^{-1}$ × $min^{-1}$.

These ACSM formulas can be implemented with the training software for medical equipment:

$$\text{Walking: } VO_2 = (1.66 \times S) + (0.30 \times S \times G) + 3.5 \tag{1.3}$$

$$\text{Running: } VO_2 = (3.33 \times S) + (0.15 \times S \times G) + 3.5 \tag{1.4}$$

where: $VO_2$, oxygen uptake expressed in mL × $kg^{-1}$ × $min^{-1}$; S, speed in km/h; G, grade expressed in %; 3.5, resting component in mL × $kg^{-1}$ × $min^{-1}$.

Loss of precision in meticulous VO2peak estimation on a treadmill occurs due to the following reasons [4, 5]:

- gripping the handrails by the tested subject
- the patient's age and clinical condition
- test protocol incorrectly matched the patient's exercise capacity
- failure to achieve a steady state at the peak of exercise.

Based on the CPET tests, the relationship between VO2 and the speed and incline of the treadmill was analyzed using multivariable linear regression to develop a more accurate FRIEND equation [5].

$$VO_2 = \mathrm{S} \times (0.17 + G \times 0.79) + 3.5 \tag{1.5}$$

where: $VO_2$, oxygen uptake expressed in mL × kg$^{-1}$ × min$^{-1}$; S, speed in m/min; G, grade expressed in decimal format (eg, 10% = 0.10); 3.5, resting component in mL × kg$^{-1}$ × min$^{-1}$.

The FRIEND equation adopted to used units is presented below:

$$VO_2 = 16.66 \times \mathrm{S} \times (0.17 + G \times 0.0079) + 3.5 \tag{1.6}$$

where: $VO_2$, oxygen uptake expressed in mL × kg$^{-1}$ × min$^{-1}$; S, speed in km/h; G, grade expressed in %; 3.5, resting component in mL × kg$^{-1}$ × min$^{-1}$.

In the study group of almost eight thousand healthy people, the FRIEND equation predicted VO2max with a much smaller error than the ACSM formula, 5.1 ± 18.3% vs 21.4 ± 24.9%, respectively [5]. Similar results were obtained in a group of 293 coronary artery disease (CAD) patients, in whom the predicted VO2peak value calculated adopting the FRIEND equation did not vary from the CPET measurement. VO2 was estimated by the ACSM formula in the study group of patients, and the measurement was substantially overestimated by 20% in men and 16% in women [6].

Estimating VO2 in a cycle ergometer test is usually done using the ACSM equation [3] presented below:

$$VO_2 = \frac{1.8 \times WR}{BM} + 3.5 + 3.5 \tag{1.7}$$

where: $VO_2$, oxygen uptake expressed in mL × kg$^{-1}$ × min$^{-1}$; WR, work rate in kg × m × min$^{-1}$; BM, body mass in kg; 3.5, resting component in mL × kg$^{-1}$ × min$^{-1}$; 3.5, horizontal component in mL × kg$^{-1}$ × min$^{-1}$.

The ACSM cycle ergometer equation adapted to work rate in watts is presented below:

$$VO_2 = 1.8 \times \frac{WR \times 6.12}{BM} + 7.0$$

$$VO_2 = 11.0 \times \frac{WR}{BM} + 7.0 \tag{1.8}$$

where: $VO_2$, oxygen uptake expressed in mL × kg$^{-1}$ × min$^{-1}$; WR, work rate in watts; BM, body mass in kg; 7.0, resting + horizontal component in mL × kg$^{-1}$ × min$^{-1}$.

The data of a few thousand CPET studies on a cycle ergometer recorded in the FRIEND register were used to develop a new regression equation for estimating VO2 based on the tolerable peak work rate in watts, body weight, and gender. The new FRIEND-ergometry equation allowed estimating VO2 with an average difference of 0.5% ± 0.11 compared to 15.46% ± 0.13 error in the ACSM equation estimation. The equations for men, women, and the general population are presented below [7]:

$$\text{Men: } VO_2 = 1.76 \times \frac{(WR \times 6.12)}{BM} + 3.5 \tag{1.9}$$

$$\text{Women: } VO_2 = 1.65 \times \frac{(WR \times 6.12)}{BM} + 3.5 \tag{1.10}$$

$$\text{Non-gender specific } VO_2 = 1.74 \times \frac{(WR \times 6.12)}{BM} + 3.5 \tag{1.11}$$

where: $VO_2$, oxygen uptake expressed in mL × kg$^{-1}$ × min$^{-1}$; WR, work rate in watts; BM, body mass in kg; 3.5, resting component expressed in mL × kg$^{-1}$ × min$^{-1}$.

The MET parameter indicates the amount of oxygen consumed while resting in sitting position, equating to 3.5 mL × kg$^{-1}$ × min$^{-1}$. This can be calculated according to formula [8]:

$$MET = VO_2\left(mL \times kg^{-1} \times min^{-1}\right)/3.5\left(ml \times kg^{-1} \times min^{-1}\right) \tag{1.12}$$

Exercise testing can be performed as maximal or submaximal depending on the indication, clinical picture, and time from cardiac events.

In CR practice, two protocols are used in exercise testing, such as:

- a graded protocol—with constant load steps lasting from one to three minutes,
- and ramp protocols with a constant increase in load with or without a short constant warm-up phase.

Ramp exercise protocols are recommended due to linear correlation with hemodynamic parameters.

A list of types of exercise tests routinely used in cardiac rehabilitation is presented in Table 1.1.

**Table 1.1** Types of exercise tests used in cardiac rehabilitation

| Division criterion | Types of exercise tests |
|---|---|
| Type of ergometer mode | – on a treadmill<br>– on a cycle ergometer |
| Measurement of oxygen consumption | – ExT—indirect measurement based on mechanical parameters<br>– CPET—direct measurement |
| Peak load | – maximal (limited by symptoms)<br>– submaximal (predefined by HR, RPE) |
| Type of protocols | – graded test<br>– ramp test |

ExT, electrocardiographic exercise test; CPET, cardio-pulmonary exercise test: HR, heart rate; RPE, rate of perceived exertion

## 1.1.1 Indications to Exercise Test

Indications of exercise tests include [9]:

- confirmation of coronary artery disease in patients with angina or symptom equivalents
- evaluation of the anatomical and functional progression of coronary artery disease
- assessment of physical capacity
- evaluation of exercise-related symptoms, chronotropic competence, and arrhythmia
- exercise-related response to implanted device therapy or to medical treatment.

Exercise tests in cardiac rehabilitation are specifically targeting (Table 1.2):

- entry risk stratification for cardiac events during exercise based on functional capacity, coronary reserve, hemodynamic response (BP, HR), and the presence of exertional arrhythmias

**Table 1.2** Cardiac rehabilitation entry and discharge exercise tests indications

| Entry exercise test | Discharge exercise test |
|---|---|
| Assessment of treatment after acute phase of cardiac disease | Functional capacity improvement after exercise program |
| Selection of patients with positive test for urgent procedure | Prognostic assessment |
| Risk stratification for exercise-induced cardiac events | Selection of patients for invasive procedure |
| Qualification to specific training model with training load determination | Return to work assessment |
| | Recreational activity planning and qualification to the next cardiac rehabilitation phase |

**Table 1.3** Absolute and relative contraindications to exercise test entail [10]

| Absolute | Relative |
|---|---|
| Myocardial infarction (MI) within 4 days | Left main coronary artery disease or its equivalent |
| Ongoing unstable angina | Tachyarrhythmias with uncontrolled ventricular rate |
| Decompensated heart failure | High-degree atrioventricular block without a pacemaker |
| Uncontrolled cardiac arrhythmia with hemodynamic compromise | Moderate-to-severe aortic stenosis |
| Symptomatic severe aortic stenosis | Hypertrophic obstructive cardiomyopathy with high resting gradient |
| Active endocarditis | Resting blood pressure of >200/110 mm Hg |
| Acute pulmonary embolism, pulmonary infarction, or deep vein thrombosis | Significant anemia, electrolyte imbalance, and hyperthyroidism |
| Physical disability precluding safe and adequate testing | Mental impairment with limited cooperation |
| Lack of patient consent | |

- functional capacity assessment and exercise prescription upon admission to phases II or III, including choice of an adequate training model, assessment of training progression, and planning of vocational and recreational activity after discharge from the program.

### *1.1.2 Contraindications to Exercise Test*

Although exercise testing is a safe procedure, the absolute and relative contraindications exist and entail specific conditions listed in Table 1.3. If relative contraindication is present, the physician in charge decides whether the potential benefits of the information derived from the test outweigh the related risk.

Details regarding staff qualifications, laboratory requirements for exercise tests, and patient preparation are available elsewhere [11].

### *1.1.3 Individualization of the Test Protocol*

The exercise test protocol should be designed to achieve a fatigue-limited duration of 8 to 12 min. The selection of an individualized test protocol should follow the two steps illustrated in Fig. 1.1.

Therefore, definitions used in the procedure include:

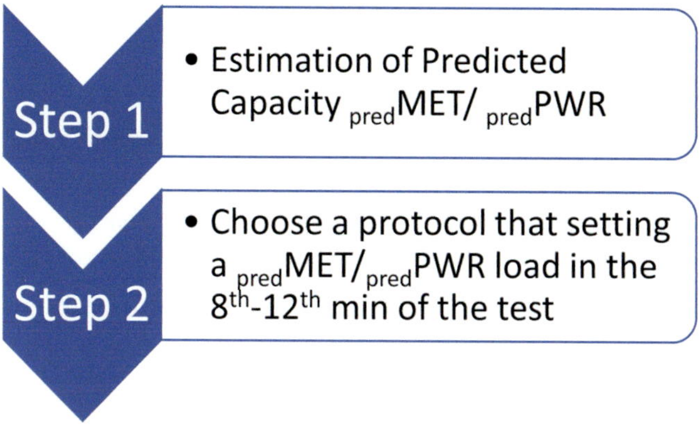


**Fig. 1.1** Two steps to selecting an individual test protocol $_{pred}$MET—predicted metabolic equivalent of task/$_{pred}$PWR—predicted peak work rate

- predicted exercise capacity (predMET)—estimated exercise capacity value in METs determined before performing the test,
- peak exercise capacity (peakMET, PWR)—physical capacity expressed in METs or peak work rate determined in an exercise test on a treadmill or bicycle ergometer,
- reference exercise capacity (refMET)—mean MET value in healthy test groups compared by age and sex (Fig. 1.2).

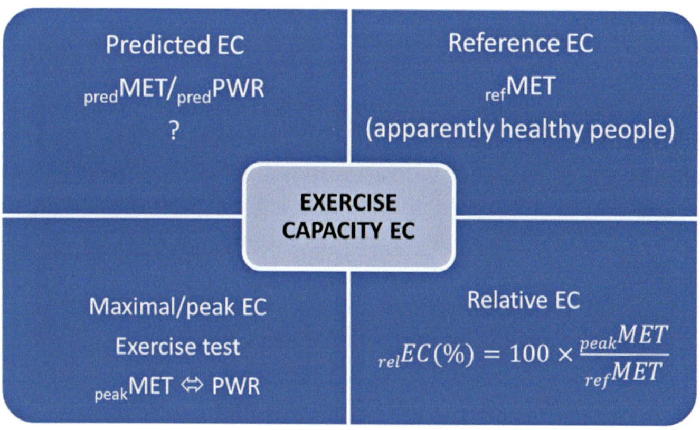


**Fig. 1.2** The term "Exercise Capacity EC" used in stress testing $_{pred}$MET, predicted metabolic equivalent of task; $_{pred}$PWR, predicted peak work rate; $_{ref}$MET, reference metabolic equivalent of task; $_{peak}$MET, peak metabolic equivalent of task; PWR, peak work rate; $_{rel}$EC, relative exercise capacity

### 1.1.4 *Estimation of the Patient's Predicted Capacity—Step I*

- Pretest clinical evaluation
- Assessment of predicted cardiorespiratory fitness (CRF) using:
  - The Veterans Specific Activity Questionnaire VSAQ
  - regression equation for the apparently healthy population
  - regression equation for patients with cardio-vascular diseases (CVD).
- Determination of the patient's fitness category in relation to the reference exercise capacity.

#### Pretest Clinical Evaluation

Pretest clinical evaluation entails medical history and examination, analysis of medical records, electrocardiogram (ECG), echocardiography (ECHO), and 24-h ambulatory ECG. The medical evaluation incorporates the assessment of the diagnosis, the symptoms declared, the presence of cardiovascular risk factors and comorbidities, and the medical treatment. In the case of implanted cardiac electrical devices, the device characteristics should be known. Patients after cardiac surgery require careful assessment of the postoperative wound healing status and the presence of pericardial effusion by echocardiography examination.

#### Veterans Specific Activity Questionnaire

One of the most used questionnaires in physical performance assessment is the Veterans Specific Activity Questionnaire (VSAQ), which contains 13 groups of daily, recreational, and sports activities with corresponding MET values [12]. The subject is asked to indicate a group of activities that can be performed with minimal or no symptoms of exercise intolerance, such as fatigue, shortness of breath, or chest discomfort.

The predicted MET value correlated with the subject's age and VSAQ (MET) can be calculated using the following regression equation [12]:

$$predMET = 4.7 + 0.97 \times VSAQ(MET) - 0.06 \times A \tag{1.13}$$

where: predMET, predicted metabolic equivalent of task; VSAQ (METs), metabolic equivalent of task estimated from the Veterans Specific Activity Questionnaire; A, age (years).

#### Assessment of Predicted CRF in an Apparently Healthy Population

Based on 1863 CPET performed on a treadmill according to the Bruce protocol at National Aeronautics and Space Administration/Johnson Space Centre (NASA) employees aged 20–70 years, elaborated multiple linear regression models for estimation of VO2max correlated with several factors, including gender, age, body mass index, resting heart rate, and self-reported physical activity (SR-PA) score [13].

The five-level SR-PA classification with corresponding score values and the regression equation for the predicted MET calculation can be found elsewhere [13].

#### Assessment of Predicted CRF in CVD Patients

The predicted VO2peak in patients with cardiovascular diseases (CVD) can be estimated using a regression equation developed based on maximal cardiopulmonary exercise testing (CPET) results from 15,997 patients. These patients had an average age of $63.1 \pm 10.4$ years, with 83% being men. The data was collected from individuals who had undergone percutaneous coronary intervention (PCI), experienced a myocardial infarction (MI), underwent coronary artery bypass graft (CABG) surgery, or suffered from heart failure (HF) as part of the FRIEND registry.

In this group of patients, VO2peak depended on several factors, including age, sex, height, weight, exercise mode, and the specific CVD diagnosis. The following stepwise multiple regression equation represents this relationship [14]:

$$\begin{aligned} VO_{2peak} &= 16.18 - (0.22 \times A) + (3.63 \times sex) + (0.14 \times H) - (0.12 \times BM) \\ &\quad + (3.62 \times mode) - (2.70 \times CABG) - (0.31 \times MI) \\ &\quad + (0.37 \times PCI) - (4.47 \times HF) \end{aligned} \tag{1.14}$$

where: $VO_{2peak}$, predicted peak oxygen uptake ($mL \times kg^{-1} \times min^{-1}$); A, age (years); sex, male: sex = 1, female: sex = 0; H, height (cm); BM, body mass (kg); mode, treadmill: mode = 1, cycle: mode = 0; CABG, coronary artery bypass graft surgery: yes CABG = 1, no CABG = 0; MI, myocardial infarction: yes MI = 1, no MI = 0; PCI, percutaneous coronary intervention: yes PCI = 1, no PCI = 0; HF, heart failure: yes HF = 1, no HF = 0.

### 1.1.5 Reference Values of Physical Fitness for Men and Women

Functional capacity depends on age, sex, habitual activity, clinical status, genetic factors, and mode of exercise (treadmill or bicycle ergometer).

The subject's predicted exercise capacity (predMET) should be compared to the reference value in order to calculate relative exercise capacity (relEC [%]) and determine the predicted fitness category.

Based on maximal treadmill exercise tests performed on 1,388 men with an average age of 57 years (21–89 years) diagnosed with coronary heart disease, a regression equation was developed to calculate the reference exercise capacity (refMET) according to the following formula [15]:

$$Men : refMET = 18.0 - 0.15 \times A \tag{1.15}$$

where: refMET, reference metabolic equivalent of task for age; A, age (years).

In a study involving 5,721 asymptomatic women, reference exercise capacity values were determined using symptom-limited treadmill exercise tests. The linear regression equation for calculating the reference MET based on age is shown below [16]:

$$Women : refMET = 14.7 - 0.13 \times A \tag{1.16}$$

where: refMET, reference metabolic equivalent of task for age; A, age (years).

In both cited articles, physical capacity (VO2peak) was estimated according to the ACSM equation (Eq. 1.1) and converted into MET units (Eq. 1.11).

Reference MET was calculated according to the above regression equations for men and women, which present average values for age.

To determine the expected physical capacity class, the expected relative capacity should be calculated according to the following formula. In both cited articles, physical capacity (VO2peak) was estimated according to the ACSM equation (Eq. 1.1) and converted into MET units (Eq. 1.11).

Reference MET was calculated according to the above regression equations for men and women, which present average values for age.

The expected relative capacity should be calculated using the following formula to determine the expected physical capacity class.

$$predictedrelEC = 100 \times \frac{predMET}{refMET} \tag{1.17}$$

where: relEC, relative exercise capacity (%); predMET, estimated predicted metabolic equivalent of task; refME, calculated reference metabolic equivalent of yak for subject age.

### 1.1.6 *Fitness Categories for Estimated MET*

The seven-stage fitness category proposed by the ACSM allows for calculating class relative efficiency by determining the ratio of VO2 in individual classes to the VO2 in the average class.

The fitness category in relative exercise capacity classification consists of the following classes [17]:

- Excellent—143% to 252% (mean 165%)
- Good—120% to 142% (mean 130%)
- Above average—108% to 119% (mean 115%)
- Average—96% to 107% (mean 100%)
- Below average—85% to 95% (mean 90%)
- Poor—75% to 84% (mean 80%)
- Very poor—39% to 74% (mean 60%).

The first step of the procedure for selecting an individualized test protocol ends with determining the predicted exercise capacity in MET and its relative capacity class.

### 1.1.7 *Test Protocol Selection—Step II*

In the second step of the procedure, a ready-made graded or ramp protocol is selected that sets the predicted load in the range of the 8th–12th minute of the test or from a set of individual ramp protocols that set predMET to the 10th minute of the test.

In the scientific literature, it is possible to find descriptions of many protocols and their comparisons [10]. However, their selection should be guided by the set of loads in the optimal time to achieve the target fatigue in the examined patient.

#### Graded Treadmill Test Protocols

Table 1.3 shows the MET load ranges at the optimal completion time for selected graded exercise testing protocols (Table 1.4).

The protocols Weber, modified Bruce, and Naughton in the above list are intended for the assessment of patients with low performance (e.g., 4–7 METs). The next three protocols, Cornell, Lehmann, and ACIP, cover the load range of 7–10 METs optimal for completing the test. More aggressive protocols such as Bruce or Ellestad should be selected in exercise testing of patients with good performance above 10 METs.

A practical solution for selecting a graded test protocol is, for example, the use of one protocol from each of the three groups mentioned, according to the predicted capacity ($_{pred}$MET) of the tested patient:

**Table 1.4** Workload in 8–12th min in graded exercise tests (adopted with permission from Ref. [19])

| Test protocol | Workload in the 8–12th min of the test (MET)* | Workload in the 8–12th min of the test (MET)** |
|---|---|---|
| Weber [18] | 4.5–6.4 | 4.5–6.3 |
| Bruce modified [19] | 4.6–7.1 | 4.2–6.1 |
| Naughton [10] | 5.4–6.9 | 4.9–5.7 |
| Cornell [20] | 7.1–10.2 | 6.1–8.3 |
| Lehmann [21] | 6.6–8.6 | 6.0–7.2 |
| ACIP [22] | 7.6–10.5 | 6.8–8.1 |
| Ellestad [23] | 9.6–11.7 | 8.6–10.5 |
| Bruce [24] | 10.2–13.5 | 8.3–10.5 |

ACIP, Asymptomatic Cardiac Ischemia Pilot protocol
(MET)*, metabolic equivalent of task calculated according to the American College of Sports Medicine equation [3]
(MET)**, metabolic equivalent of task calculated according to Fitness Registry and the Importance of Exercise National Database equation [5]

- Naughton or modified Bruce protocols
- Cornell protocol
- Bruce protocol.

The Naughton protocol is intended to increase the load by approximately 1 MET for each 2—min stage. This protocol is recommended for patients with low predicted functional capacity or low gait speed (Fig. 1.3).

The modified Bruce protocol is dedicated for individuals with predicted functional capacity of 4–7 METs. Protocol incorporates two initial stages at lower load (Fig. 1.4).

Cornell protocol incorporates 2-min stages and intermediate loads between standard Bruce stages, evoking mild load increase (Fig. 1.5).

The protocol developed and published in 1963 by Bruce [24] allowed for the standardization and comparison of test results and is the basis for many diagnostic and prognostic tests. Bruce Protocol is a standard protocol (Fig. 1.6) and is the most popular graded treadmill protocol. However, due to large load increments, it should be conducted only for individuals with a good predicted functional capacity (i.e., above 10 METs).

## Ramp Protocols

Ramp tests demonstrate a linear correlation between the load and the maximal oxygen consumption and are better tolerated by patients [25, 26].

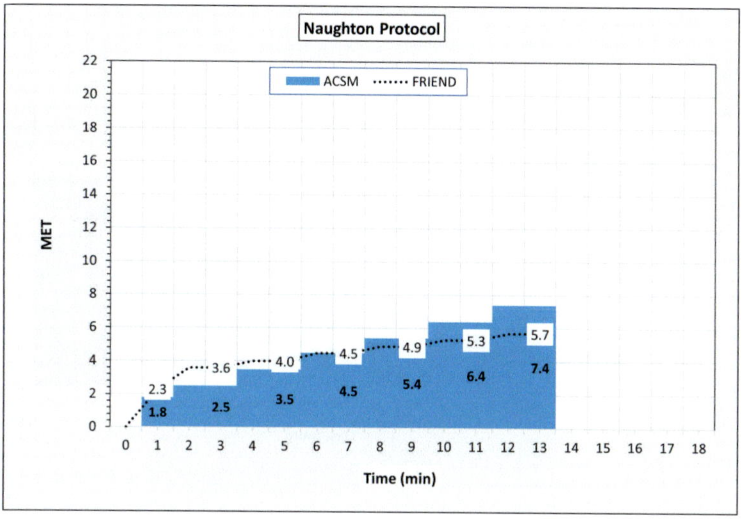


**Fig. 1.3** Naughton protocol [10] ACSM, metabolic equivalent of task estimated according to the American College of Sports Medicine equation [3]; FRIEND, metabolic equivalent of task estimated according to Fitness Registry and the Importance of Exercise National Database equation [5]

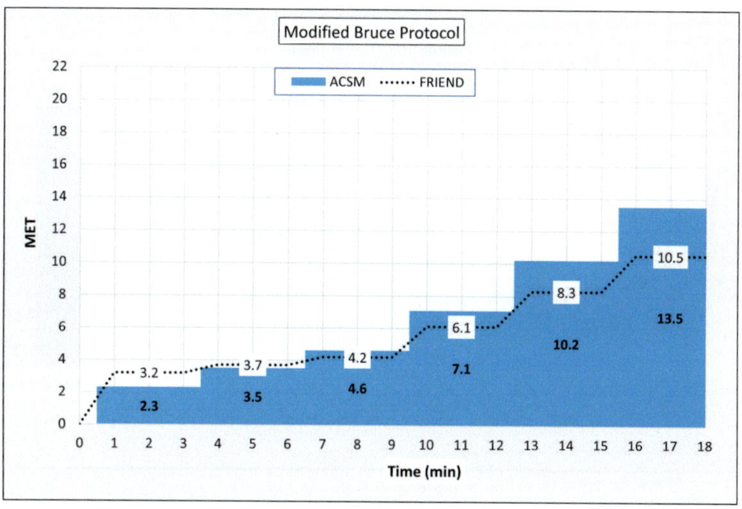


**Fig. 1.4** Modified Bruce protocol [19]. ACSM, metabolic equivalent of task estimated according to the American College of Sports Medicine equation; FRIEND, metabolic equivalent of task estimated according to Fitness Registry and the Importance of Exercise National Database equation

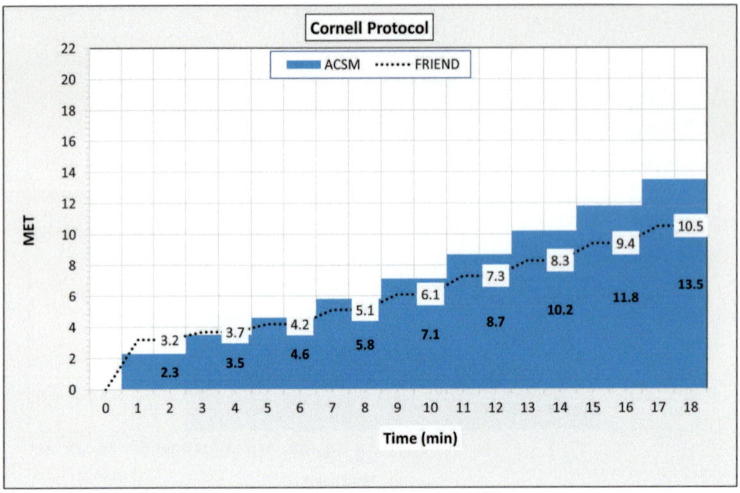


**Fig. 1.5** Cornell protocol [20]. ACSM, metabolic equivalent of task estimated according to the American College of Sports Medicine equation [3]; FRIEND, metabolic equivalent of task estimated according to Fitness Registry and the Importance of Exercise National Database equation [5]

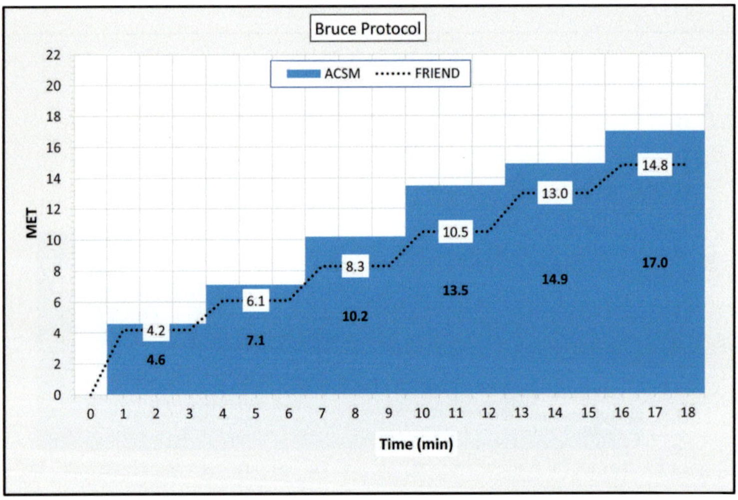


**Fig. 1.6** Bruce graded protocol [24]. ACSM, metabolic equivalent of task estimated according to the American College of Sports Medicine equation [3]; FRIEND, metabolic equivalent of task estimated according to Fitness Registry and the Importance of Exercise National Database equation [5]

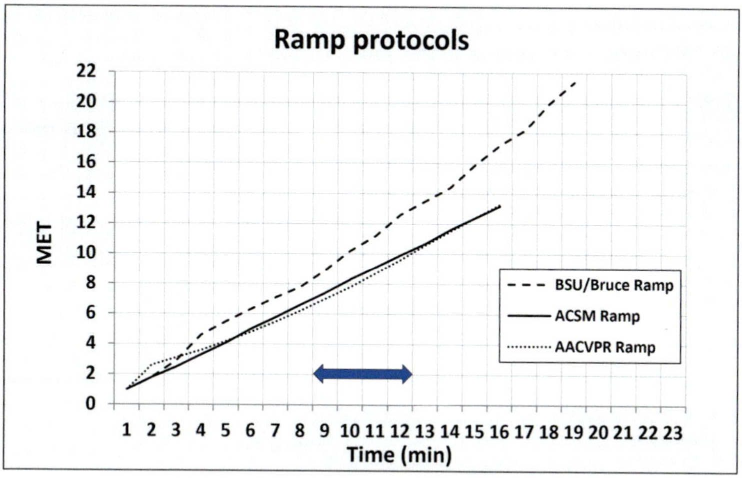


**Fig. 1.7** Treadmill ramp protocols. BSU/Bruce Ramp, Ball State University/Bruce Ramp Protocol [26]. ACSM Ramp, American College of Sports Medicine Ramp Protocol [28], AACVPR Ramp, American Association of Cardiovascular and Pulmonary Rehabilitation Ramp Protocol [27]

Features of the treadmill ramp protocol introduced by Jonathan Myers [4], include:

- a gentle start of the load as a warm-up
- constant steady increase in load
- linear hemodynamic and ventilation adaptation
- better load tolerance
- more accurate capacity determination
- the possibility of individualizing the protocol.

The proposed ramp tests by AACVPR [27] and ACSM [28] impose almost identical loads (i.e., ~7–10 METs) in the range of 8th–12th min of the test, while the BSU/Bruce Ramp Test [26] set a load of 9–13.5 METs, respectively (Fig. 1.7).

## Individual Ramp Protocols

Individualized ramp test protocols allow for meticulous load delivery and termination between the 8th and 12th min of the test [29].

There are three approaches to selecting a customized ramp protocol, listed below:

- performing a graded preliminary test with determination of peak capacity (METpeak) and programming a ramp protocol with METpeak in 10th min of the test,
- using specialized software integrated with the exercise testing system, which, based on the predicted performance (predMET), creates a protocol with predMET in 10th min of the tests,

- individual selection of a ramp protocol from a set of protocols with different loads in the 10th minute, according to the subject's expected capacity.

Rybicki [30] proposed a customized set of individualized ramp protocols with the number added according to MET values (calculated according to the ACSM equation) attained exactly in the tenth minute of the test. Tables 1.5, 1.6, and 1.7 present individualized ramp test protocols.

MET loads estimated according to the ACSM are displayed graphically, and Fig. 1.8 shows MET loads estimated according to the ACSM and FRIEND equations in individualized ramp protocols.

**Table 1.5** Individualized Ramp 5.0, Ramp 6.5 protocols according to Rybicki (adopted with permission from Ref. [19])

| Time (min) | Ramp 5.0 | | | | Ramp 6.5 | | | |
|---|---|---|---|---|---|---|---|---|
| | S (km/h) | G (%) | MET ACSM | MET FRIEND | S (km/h) | G (%) | MET ACSM | MET FRIEND |
| Start | 1.8 | 1.8 | 2.1 | 2.5 | 2.0 | 2.0 | 2.3 | 2.7 |
| 1 | 1.8 | 1.8 | 2.1 | 2.5 | 2.0 | 2.0 | 2.3 | 2.7 |
| 2 | 1.9 | 3.2 | 2.4 | 2.8 | 2.2 | 3.8 | 2.7 | 3.1 |
| 2 | 2.0 | 4.4 | 2.7 | 3.0 | 2.4 | 5.3 | 3.2 | 3.4 |
| 4 | 2.2 | 5.5 | 3.1 | 3.2 | 2.6 | 6.6 | 3.7 | 3.7 |
| 5 | 2.3 | 6.5 | 3.4 | 3.4 | 2.8 | 7.7 | 4.2 | 4.0 |
| 6 | 2.4 | 7.3 | 3.7 | 3.7 | 3.0 | 8.7 | 4.6 | 4.4 |
| 7 | 2.6 | 8.1 | 4.0 | 3.9 | 3.2 | 9.5 | 5.1 | 4.7 |
| 8 | 2.7 | 8.8 | 4.4 | 4.1 | 3.4 | 10.2 | 5.6 | 5.0 |
| 9 | 2.9 | 9.4 | 4.7 | 4.3 | 3.6 | 10.9 | 6.0 | 5.4 |
| 10 | 3.0 | 10.0 | 5.0 | 4.6 | 3.8 | 11.5 | 6.5 | 5.7 |
| 11 | 3.1 | 10.5 | 5.3 | 4.8 | 4.0 | 12.0 | 7.0 | 6.0 |
| 12 | 3.3 | 10.9 | 5.6 | 5.0 | 4.2 | 12.5 | 7.4 | 6.3 |
| 13 | 3.4 | 11.4 | 6.0 | 5.2 | 4.4 | 12.9 | 7.9 | 6.7 |
| 14 | 3.6 | 11.7 | 6.3 | 5.5 | 4.6 | 13.3 | 8.4 | 7.0 |
| 15 | 3.7 | 12.1 | 6.6 | 5.7 | 4.8 | 13.7 | 8.8 | 7.3 |
| 16 | 3.8 | 12.5 | 6.9 | 5.9 | 5.0 | 14.0 | 9.3 | 7.6 |
| 17 | 4.0 | 12.8 | 7.3 | 6.1 | 5.2 | 14.3 | 9.8 | 8.0 |
| 18 | 4.1 | 13.1 | 7.6 | 6.4 | 5.4 | 14.6 | 10.3 | 8.3 |

S, speed; G, grade; MET ACSM, metabolic equivalent of task estimated according to American College of Sports Medicine equation (Eq. 1.3); MET FRIEND, metabolic equivalent of task estimated according to Fitness Registry and the Importance of Exercise National Database equation (Eq. 1.6)

**Table 1.6** Individualized Ramp 8.0, Ramp 10.0 protocols according to Rybicki (adopted with permission from Ref. [19])

| Time (min) | Ramp 8.0 | | | | Ramp 10.0 | | | |
|---|---|---|---|---|---|---|---|---|
| | S (km/h) | G (%) | MET ACSM | MET FRIEND | S (km/h) | G (%) | MET ACSM | MET FRIEND |
| Start | 2.2 | 2.2 | 2.5 | 3.0 | 2.5 | 2.5 | 2.7 | 3.3 |
| 1 | 2.2 | 2.2 | 2.5 | 3.0 | 2.5 | 2.5 | 2.7 | 3.3 |
| 2 | 2.5 | 4.3 | 3.1 | 3.4 | 2.8 | 4.9 | 3.5 | 3.8 |
| 2 | 2.7 | 6.1 | 3.7 | 3.8 | 3.1 | 6.9 | 4.3 | 4.3 |
| 4 | 3.0 | 7.5 | 4.3 | 4.2 | 3.4 | 8.5 | 5.1 | 4.9 |
| 5 | 3.2 | 8.7 | 4.9 | 4.6 | 3.8 | 9.8 | 6.0 | 5.4 |
| 6 | 3.5 | 9.7 | 5.5 | 5.1 | 4.1 | 11.0 | 6.8 | 6.0 |
| 7 | 3.7 | 10.6 | 6.2 | 5.5 | 4.4 | 11.9 | 7.6 | 6.5 |
| 8 | 4.0 | 11.4 | 6.8 | 5.9 | 4.7 | 12.8 | 8.4 | 7.1 |
| 9 | 4.2 | 12.1 | 7.4 | 6.3 | 5.0 | 13.5 | 9.2 | 7.6 |
| 10 | 4.5 | 12.7 | 8.0 | 6.8 | 5.3 | 14.1 | 10.0 | 8.2 |
| 11 | 4.7 | 13.3 | 8.6 | 7.2 | 5.6 | 14.7 | 10.8 | 8.7 |
| 12 | 5.0 | 13.7 | 9.2 | 7.6 | 6.0 | 15.2 | 11.6 | 9.2 |
| 13 | 5.2 | 14.2 | 9.8 | 8.0 | 6.3 | 15.7 | 12.4 | 9.8 |
| 14 | 5.5 | 14.6 | 10.5 | 8.4 | 6.6 | 16.1 | 13.2 | 10.3 |
| 15 | 5.7 | 15.0 | 11.1 | 8.9 | 6.9 | 16.5 | 14.0 | 10.9 |
| 16 | 6.0 | 15.3 | 11.7 | 9.3 | 7.2 | 16.8 | 14.8 | 11.4 |
| 17 | 6.2 | 15.6 | 12.3 | 9.7 | 7.5 | 17.1 | 15.7 | 12.0 |
| 18 | 6.5 | 15.9 | 12.9 | 10.1 | 7.9 | 17.4 | 16.5 | 12.5 |

S, speed; G, grade; MET ACSM, metabolic equivalent of task estimated according to American College of Sports Medicine equation (Eq. 1.3); MET FRIEND, metabolic equivalent of task estimated according to Fitness Registry and the Importance of Exercise National Database equation (Eq. 1.6)

## Cycle Ramp Protocols

Traditional cycle ergometer-graded exercise test protocols typically include a 1-min warm-up, which may involve a load of 10 to 20 W or no load. After the warm-up, the initial load is set at 25 W for inactive individuals or those weighing less than 70 kg, while active individuals or those with a weight more than 70 kg start at 50 Watts. The load is increased by 25 Watts every 2–3 min [31].

Currently, available ramp tests allow their selection depending on the capacity of the tested patient and linear load increase [32, 33] (Table 1.8).

The load advancement during ramp tests on a cycle ergometer is displayed in Fig. 1.9.

**Table 1.7** Individualized Ramp 12.0, Ramp 14.0 protocols according to Rybicki (adopted with permission from Ref. [19])

| Time (min) | Ramp 12.0 | | | | Ramp 14.0 | | | |
|---|---|---|---|---|---|---|---|---|
| | S (km/h) | G (%) | MET ACSM | MET FRIEND | S (km/h) | G (%) | MET ACSM | MET FRIEND |
| Start | 2.8 | 2.8 | 3.0 | 3.6 | 3.1 | 3.1 | 3.3 | 3.9 |
| 1 | 2.8 | 2.8 | 3.0 | 3.6 | 3.1 | 3.1 | 3.3 | 3.9 |
| 2 | 3.1 | 5.7 | 4.0 | 4.2 | 3.4 | 6.3 | 4.5 | 4.6 |
| 2 | 3.4 | 8.2 | 5.0 | 4.8 | 3.8 | 8.9 | 5.7 | 5.3 |
| 4 | 3.7 | 10.2 | 6.0 | 5.4 | 4.1 | 11.1 | 6.9 | 6.0 |
| 5 | 4.0 | 11.9 | 7.0 | 6.0 | 4.4 | 13.0 | 8.1 | 6.8 |
| 6 | 4.3 | 13.4 | 8.0 | 6.7 | 4.8 | 14.6 | 9.2 | 7.5 |
| 7 | 4.6 | 14.6 | 9.0 | 7.3 | 5.1 | 16.0 | 10.4 | 8.2 |
| 8 | 4.9 | 15.8 | 10.0 | 7.9 | 5.4 | 17.2 | 11.6 | 8.9 |
| 9 | 5.2 | 16.8 | 11.0 | 8.5 | 5.8 | 18.3 | 12.8 | 9.7 |
| 10 | 5.5 | 17.6 | 12.0 | 9.1 | 6.1 | 19.2 | 14.0 | 10.4 |
| 11 | 5.8 | 18.4 | 13.0 | 9.8 | 6.4 | 20.1 | 15.2 | 11.1 |
| 12 | 6.1 | 19.2 | 14.0 | 10.4 | 6.8 | 20.9 | 16.4 | 11.8 |
| 13 | 6.4 | 19.8 | 15.0 | 11.0 | 7.1 | 21.6 | 17.6 | 12.5 |
| 14 | 6.7 | 20.4 | 16.0 | 11.6 | 7.5 | 22.2 | 18.8 | 13.3 |
| 15 | 7.0 | 20.9 | 17.0 | 12.3 | 7.8 | 22.8 | 19.9 | 14.0 |

S, speed; G, grade; MET ACSM, metabolic equivalent of task estimated according to American College of Sports Medicine equation (Eq. 1.3); MET FRIEND, metabolic equivalent of task estimated according to Fitness Registry and the Importance of Exercise National Database equation (Eq. 1.6)

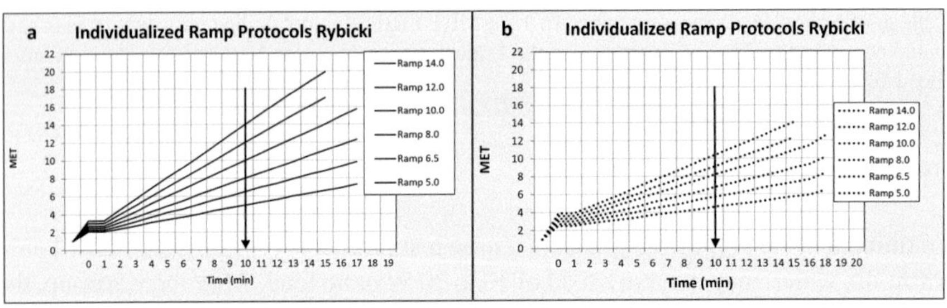


**Fig. 1.8** **a** MET load course in the protocols: Ramp 5.0, Ramp 6.5, Ramp 8.0, Ramp 10.0, Ramp 12.0 and Ramp 14.0; values estimated according to American College of Sports Medicine equation (Eq. 1.3); **b** values estimated according to the Fitness Registry and the Importance of Exercise National Database equation (Eq. 1.6)

**Table 1.8** A set of ramp tests on a cycle ergometer

| Cycle ramp protocols | Predicted capacity (MET) | Warm-up 1 min (W) | Workload increment per min (W) | Workload in 8–10–12th min (W) |
|---|---|---|---|---|
| C-Ramp 5 | 3.6–5,1 | 5 | 5 | 40–50–60 |
| C-Ramp 10 | 5.1–8.3 | 10 | 10 | 80–100–120 |
| C-Ramp 15 | 6.7–10.1 | 15 | 15 | 120–150–180 |
| C-Ramp 20 | 8.3–14,8 | 20 | 20 | 160–200–240 |
| C-Ramp 25 | 9,9–17.7 | 25 | 25 | 200–250–300 |
| C-Ramp 30 | 11,4–20,9 | 30 | 30 | 240–300–360 |

Predicted capacity in MET was estimated for workload in 8th–12th min of the test for 60–80 kg patient according to American College of Sports Medicine ergometry equation [3]
W, watts

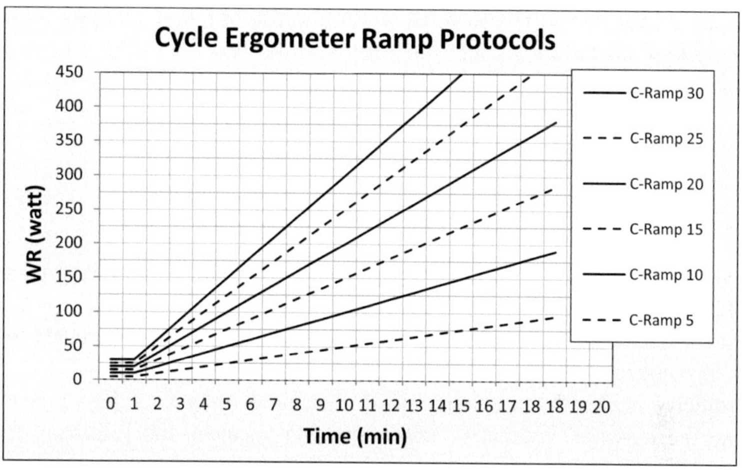


**Fig. 1.9** Cycle ergometer ramp protocols

## *1.1.8 Indications for Termination of Exercise Testing*

Full list of test termination reasons can be found in bibliography dedicated to exercise testing. The most critical reasons remain ongoing chest pain and/or ischemic ECG changes, and complex arrhythmias. The specific reason for test termination is inability to record ECG or BP monitoring.

### *1.1.9 Exercise Test Report for Cardiac Rehabilitation*

The test report should encompass several components: electrocardiographic analysis, BP and HR responses, functional capacity level, symptoms, present, and conclusions.

## 1.2 Cardiopulmonary Exercise Test in Exercise Prescription

Rationale

A cardiopulmonary exercise test (CPET) provides a comprehensive exercise physiology assessment and helps develop a tailored training prescription. CPET analyzes the exercise responses of multiple systems and serves as both a prognostic and diagnostic tool for various populations. However, the use of CPET remains limited for several reasons, including its unavailability and the time-consuming nature of its interpretation.

### *1.2.1 The Overall Procedure for Performing the Test*

Detailed requirements for CPET laboratory standards, including staff responsibilities and calibration procedures, have been described elsewhere [32, 34].

Before conducting the test, it is crucial to instruct patients appropriately to attain the best possible effort.

Spirometry evaluates lung function. A measured volume of FEV1 is used to calculate the maximal voluntary ventilation (MVV) using the following formula [35]:

$$MVV = FEV1 \times 40 \tag{1.18}$$

where: MWW, maximal voluntary ventilation (L/min); FEV1, forced expiratory volume in 1 s (L).

CPET begins with a 2–3 min recording of resting parameters, followed by an exercise phase according to the selected protocol and ending with recovery, as shown in Fig. 1.10.

The test can be performed on an ergometer or a treadmill. The bike ramp protocol should be selected according to the procedure described in Sect. 1.1.7. Starting the test with three minutes of unloaded pedaling allows the calculation of the oxygen uptake/work rate (VO2/WR) slope [36].

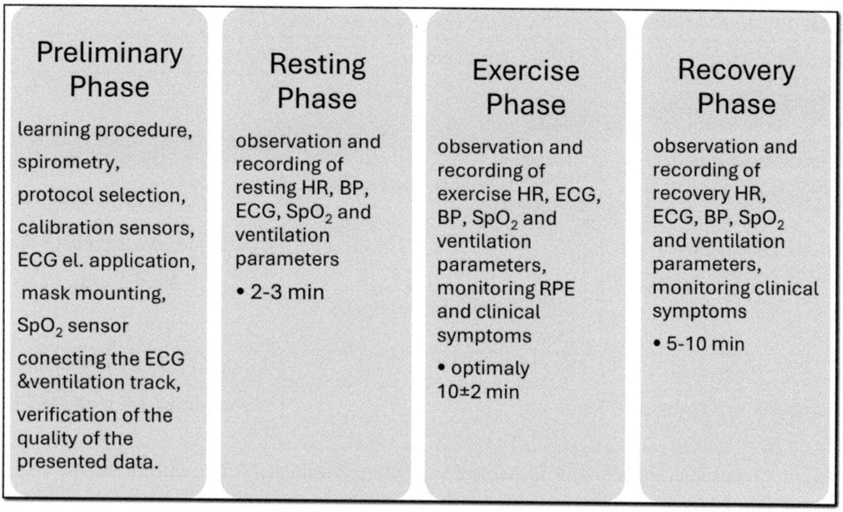


**Fig. 1.10** Phases of the cardiopulmonary exercise testing ECG, electrocardiogram; $SpO_2$, pulse oximetry; HR, heart rate; BP, blood pressure; RPE, rate of perceived exertion

## *1.2.2 Main CPET Parameters for Clinical Assessment in CR*

The parameters recommended for ExT and ventilation and gas exchange parameters should be assessed. The recognized criteria for achieving maximal effort by the subject entail [37]:

- the respiratory exchange ratio (RER) of 1.10–1.15,
- lack of further increase of VO2 and/or HR with progressive load,
- post-exercise lactate concentration ≥8 mmol/L,
- reaching RPE 18 or more (Borg 6–20 scale) or 8 or more on the Borg CR10 Scale.

Table 1.9 and Fig. 1.11 present the identification of the first and second ventilatory thresholds in CPET using a graphical trends analysis.

The key cardiopulmonary exercise testing variables in the assessment of rehabilitated patients are listed below:

**Ventilation parameters** include:

- tidal volume
- breathing frequency
- the product of the above data allows to calculate minute ventilation (VE) in L/min.

To calculate the breathing reserve (BR), the $VE_{peak}$ should be related to the maximal voluntary ventilation measured in resting spirometry.

In practice, the following three formulas are used to calculate BR:

**Table 1.9** Methods of determination of the ventilatory thresholds [34, 38]

| Method/Trends | VT1 determination | VT2 determination |
|---|---|---|
| V-slope method: $VCO_2$ versus $VO_2$ | Upward departure point of the $VCO_2$ | X |
| Equivalents method: $VE/VO_2$ ratio versus time | The nadir before an increase in $\dot{V}E/\dot{V}O_2$ with no increase in $\dot{V}E/\dot{V}CO_2$ | X |
| Equivalents method: $VE/VCO_2$ ratio versus time | X | Upward departure point of the $\dot{V}E/\dot{V}CO_2$ curve |
| $P_{ET}O_2$ versus time | Turning point before $P_{ET}O_2$ increase | X |
| $P_{ET}CO_2$ versus time | X | $P_{ET}CO_2$ curve decrease point |
| VE versus $VCO_2$ slope | X | The departure point up the $\dot{V}E$ curve |

VT1, first ventilatory threshold; VT2, second ventilatory threshold; $VCO_2$, carbon dioxide output; $VO_2$, oxygen uptake; $VE/VO_2$ ratio, ventilatory equivalent for oxygen; $VE/VCO_2$ ratio, ventilatory equivalent for carbon dioxide; $P_{ET}O_2$, end-tidal oxygen partial pressure; $P_{ET}CO_2$, end-tidal carbon dioxide partial pressure; VE, minute ventilation

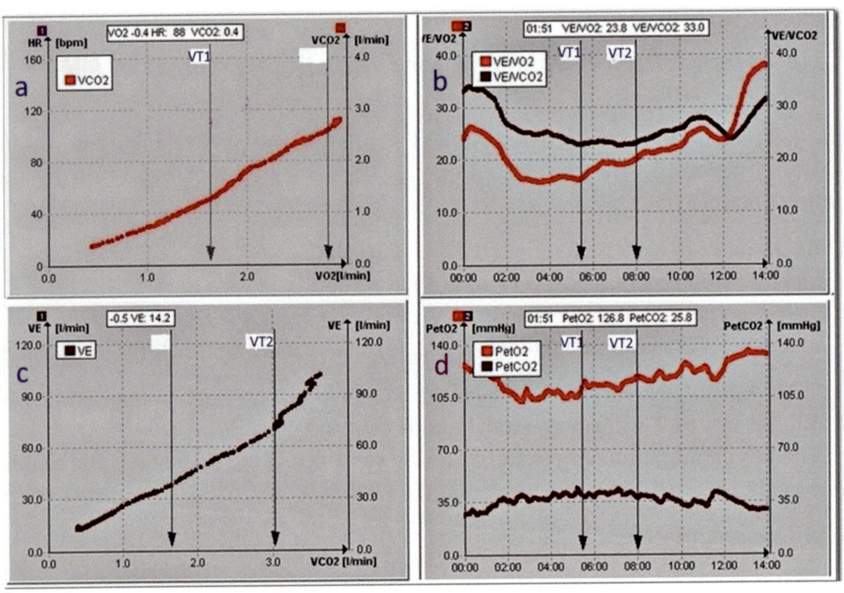


**Fig. 1.11** Graphical display of the ventilatory thresholds determination, by the authors **a** V-slope method; **b** equivalents method; **c** VE versus $VCO_2$ slope method; **d** $P_{ET}O_2$ and $P_{ET}CO_2$ versus time method—description in Table 1.9

$$\text{as a simple difference: } BR(L) = MVV - VEpeak \tag{1.19}$$

$$\text{as a real percentage reserve: } BR(\%) = 100 \times \frac{(MVV - VEpeak)}{MVV} \tag{1.20}$$

$$\text{as a simple percentage ratio: } BR(\%) = 100 \times VEpeak/MVV \tag{1.21}$$

where: BR, breathing reserve; MVV, maximal voluntary ventilation (L/min); VEpeak, ventilation on peak exercise (L/min).

**Oxygen uptake ($VO_2$)** is expressed in L × $min^{-1}$ or mL × $min^{-1}$ or as relative values:

- $VO_2$ in relation to body mass: $VO_2$ mL × $kg^{-1}$ × $min^{-1}$ (the most used clinical $VO_2$ index)
- $VO_2$ in relation to rest oxygen uptake (3.5 mL × $kg^{-1}$ × $min^{-1}$) as a metabolic equivalent of task (MET)
- a percentage of the relative exercise capacity ($_{rel}EC\%$) calculated as follows:

$$relEC(\%) = 100 \times VO_{2peak}/VO_{2ref}$$

where: $_{rel}EC$, relative exercise capacity; $VO_{2peak}$, peak oxygen uptake; $VO_{2ref}$, reference (normal) oxygen uptake.

The most frequently used are the $VO_2$ reference values, according to Wasserman/Hansen [39]; however, many publications are available elsewhere for different, larger populations studied [40–42].

**Related Parameters with $VO_2$**:

- Oxygen uptake on first and second ventilatory threshold ($VO_2$@VT1 and $VO_2$@VT2 respectively) expressed in direct measurement mL × $kg^{-1}$ × $min^{-1}$ and as a percentage of $VO_{2peak}$ or reference $VO_2$
- Oxygen pulse ($O_2$pulse) calculated as a ratio $VO_2$/HR and expressed as a peak value in mL × $min^{-1}$ × $beat^{-1}$ and presented as a trajectory (second panel according to Wasserman). It is recommended that the trajectory description indicates one of three terms [43]:
  - Continuous increase throughout exercise test
  - Early and sustained plateau
  - Decline
- Oxygen uptake in response to a simultaneous increase in workload ($\Delta VO_2/\Delta WR$) expressed in mL × $min^{-1}$ × $W^{-1}$. This parameter showing the metabolic efficiency of muscle work is presented as an average slope value analysed by linear regression method and a trajectory [36], which should be described above.

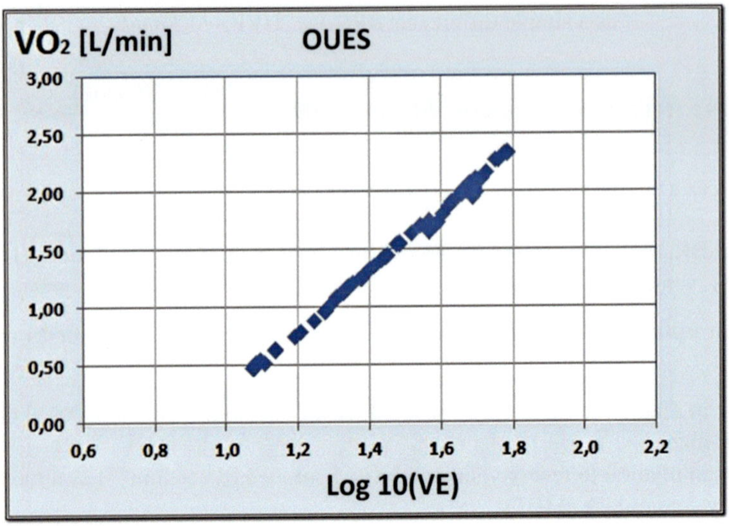


**Fig. 1.12** Oxygen uptake efficiency slope, by authors

- Oxygen uptake efficiency slope (OUES) which is the linear relation of $VO_2$ versus the logarithm of VE during incremental exercise [44] (Fig. 1.12). OUES is calculated using the linear regression equation $VO_2 = m \times \log10(VE) + b$, as a value of "m".
- Ventilatory equivalent for oxygen ($VE/VO_2$ ratio) is presented as a value and a curve used to determine VT1 (lower curve of 4th Wasserman's panel in Fig. 1.13)
- End-tidal oxygen partial pressure ($P_{ET}O_2$) expressed in mm Hg. The course of the curve allows to determine VT1 (upper waveform of 7th Wasserman's panel).

**Carbon dioxide output** at peak exercise ($VCO_{2peak}$) in $L \times min^{-1}$ or $mL \times min^{-1}$

**Related Parameters with $VCO_2$:**

- Respiratory exchange ratio $VCO_2/VO_2$ (RER),
- Ventilatory equivalent for carbon dioxide ($VE/VCO_2$ ratio) is presented as a value and a graph course used to determine VT2 (upper curve of 4th Wasserman's panel in Fig. 1.13)
- Ventilation versus carbon dioxide output slope (VE vs. $VCO_2$ slope) calculated by linear regression (6 Wasserman's plot in Fig. 1.13). VE versus $VCO_2$ is the parameter with the greatest prognostic importance in heart failure [45] with established ventilation class ranges of risk [46].
- End-tidal carbon dioxide partial pressure ($P_{ET}CO_2$) expressed in mm Hg. The course of the curve allows to determine VT2 (lower waveform of 7th Wasserman's panel in Fig. 1.13).

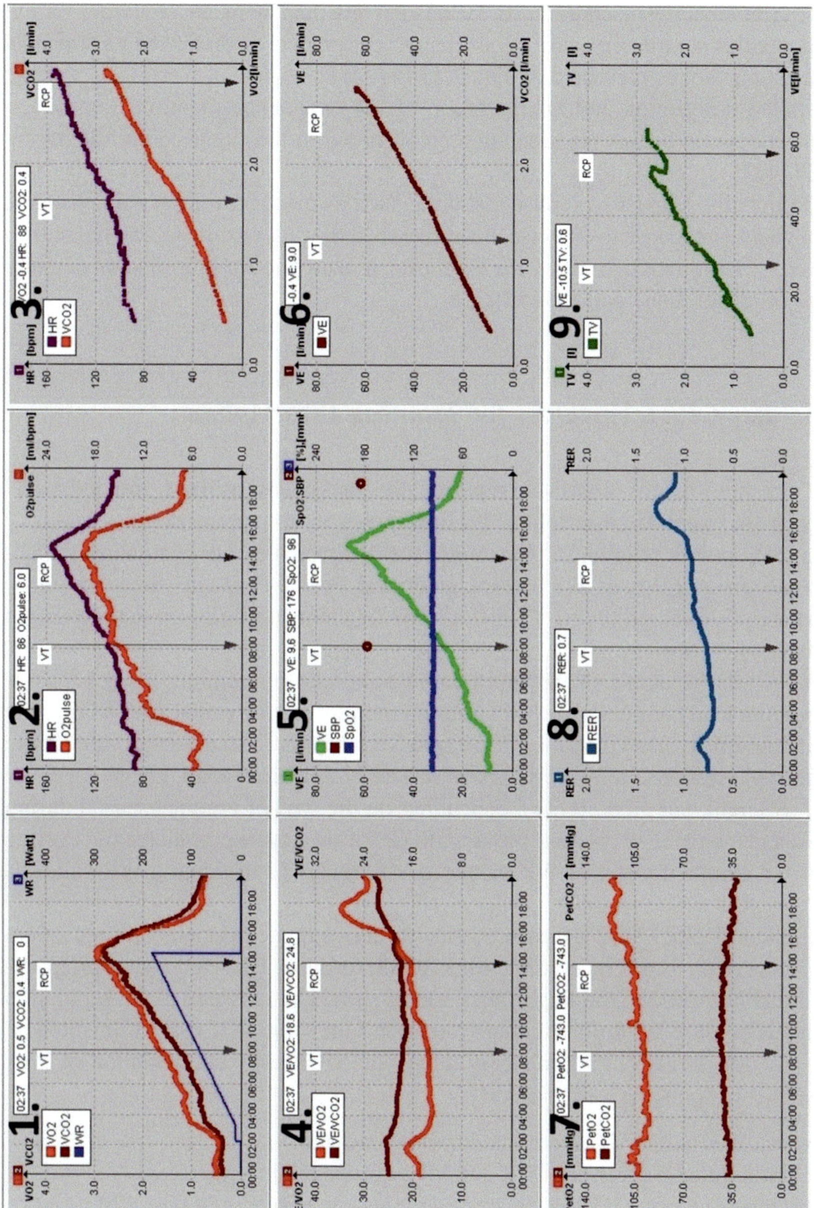


**Fig. 1.13** Wasserman's nine panels plot in submaximal ramp cardiopulmonary exercise test, by the authors

$VO_2$, $O_2$ pulse, $VE/VO_2$ ratio, $P_{ET}O_2$, RER, $VE/VCO_2$ and $P_{ET}CO_2$ variables should be presented for the resting phase, corresponding to VT1 and VT2 and at peak exercise.

The graphical representation of CPET-derived data with an overview of the cardiovascular, ventilatory and gas exchange parameters is displayed as 9 panels Wasserman's plot as presented on Fig. 1.13 [47, 48].

Detailed description and interpretation of the parameters mentioned above go beyond the scope of this position and can be found in the suggested bibliography [32, 34, 43].

Additionally, standard exercise test data, such as electrocardiographic changes, hemodynamic response to exercise, heart rate recovery after exercise, and the reason for exercise termination, improve the diagnostic and prognostic stratification of many groups of rehabilitated patients [43].

### 1.2.3 Key CPET Variables for Training Prescription

The "threshold-based" approach replaced the previously utilized "range-based" approach for exercise prescription. It allows for a more precise exercise intensity determination with relation to physiological principles, i.e., the ventilatory thresholds, than when expressed as a percentage of peak exercise capacity. Exercise intensity prescriptions based on the VT differ from indicators derived from peak exercise parameters [49, 50].

The general principle of threshold-based training programming using CPET is determining load parameters used in training monitoring correlated with VO2 at the ventilation thresholds VT1 and VT2. These variables are usually heart rate (if the chronotropic response to exercise is correct) and mechanical load, i.e., WR (W) or speed (S km/h) and grade (G %), depending on the type of ergometer used during the test and the planned exercise training. In CPET on a cycle ergometer and cycling training, a simple transfer of WR values at VT1 and VT2 to training WR values is used.

In treadmill CPET VO2 is measured directly and estimated as metabolic equivalent (eMET) based on S and G according to the ACSM formula. Therefore, when programming treadmill walking training, eMET values should be provided at the ventilation thresholds because treadmill training systems use the same VO2 estimation method.

In practice, three training intensity ranges are created by extrapolation of VT1 and VT2 corresponding variables [51]:

- low-intense (below VT1-correlated training variables),
- moderate-intense (between VT1 and VT2-corresponding variables),
- high-intense (above VT2-correlated variables) as presented in Fig. 1.14.

| Variables | Rest | VT1 | VT2 | Peak |
|---|---|---|---|---|
| $VO_2$ [mL×kg$^{-1}$×min$^{-1}$] | 3.7 | 16.4 | 27.2 | 27.4 |
| MET | 1.1 | 4.7 | 7.7 | 7.8 |
| HR [bpm] | 75 | 105 | 139 | 147 |
| % HR peak | 51 | 71 | 95 | 100 |
| % HRR | 0 | 42 | 89 | 100 |
| Treadmill set-up*<br>Speed S [km/h]<br>Grade G [%]<br>eMET | <br>0<br>0<br>1 | <br>2.6<br>8.9<br>4.2 | <br>4.6<br>12.9<br>8.2 | <br>5.0<br>13.6<br>9.1 |
| WR [W] cycle ergometer** | 0 | 82 | 117 | 125 |

**Fig. 1.14** Ramp treadmill CPET variables for determining training intensity zones using the threshold-base method for 57 years old patient after myocardial infarction VT1, first ventilatory threshold; VT2, second ventilatory threshold; $VO_2$, oxygen uptake; MET, metabolic equivalent of task; HR, heart rate; HRR, heart rate reserve; eMET, estimated metabolic equivalent of task; WR, work rate *, data from treadmill test; **, data from conversion MET to watt (see Sect. 2.3.2.1)

### *1.2.4 CPET Protocol for Cardiac Rehabilitation*

The CPET key experts developed a joint document [43, 52], containing the "Universal CPX reporting form". This document enables the registration of many parameters useful in diagnostic and prognostic stratification of several groups of cardiac and pulmonary diseases, as well as perioperative risk. At the same time, a precise methodology for measuring many CPET variables was defined.

The numerous CPET variables allow for interpreting both functional capacity and specific exertional limitations; however, most of these parameters have not been applied to an exercise prescription. Cardiac rehabilitation centers utilizing CPET developed dedicated report forms, however, a more systematic approach has been introduced recently. The Italian Alliance for Cardiovascular Rehabilitation and Prevention (ITACARE-P) published a CPET reporting form considering the need for cardiac rehabilitation [53]. The report contains the main data allowing for the clinical assessment of groups of rehabilitated patients and training intensity parameters corresponding to the ventilation thresholds, as well as data used in the range method according to the ACSM or ESC guidelines for aerobic training protocols.

## 1.3 Field Tests

Field tests are simple and low-cost methods to evaluate exercise capacity and response to treatment. They allow for submaximal testing, so they are considered valuable tools for assessing the response to the activities of daily living. However, their correlation with peak oxygen uptake is modest, and their result is strongly affected by familiarity and the patient's motivation. Thus, the usefulness of the field tests is limited to specific populations.

### *1.3.1 Six-Minute Walk Test*

The six-min walking test measures the walking distance on a flat, firm surface completed within six minutes.

The test should be performed indoors, optimally in an enclosed corridor that is seldom trafficked. As a self-paced test, it is not recommended to be performed on a treadmill due to the patient's unfamiliarity with the treadmill and significantly lower walk distance attained. As patients are walking at their own pace, the outcomes of the 6MWT should be considered complementary, and tests should not be used for exercise-related risk stratification.

The 6MWT has been traditionally used for patients in an early post-surgical period following incomplete revascularization. The 6 MWT evaluates the response for the activities of daily living and, thus, can be suitable for older people. The performance of the 6MWT requires 30 m long walking track marked every three meters. If needed, patients may use a walking stick or frame and can slow down or stop during the test. Details describing the 6MWT procedure can be found elsewhere [54].

Due to the learning curve the two tests are recommended.

Absolute contraindications for the 6MWT include recent myocardial infarction, whereas relative contraindications include resting tachycardia, high SBP, and/or DBP. The test should be terminated immediately in case of angina, marked tiredness, or other clinical instability. The test report should include distance covered, reason for termination, symptoms, recorded vital signs, and RPE. Obese individuals and older people cover shorter distances, typically below 300 m.

Enright suggested reference equations for women and men [55]:

$$\text{Women:}ref6MWD = (2{,}11 \times H) - (2{,}29 \times BM) - (5{,}78 \times A) + 667m \quad (1.21)$$

$$\text{Men:}ref6MWD = (7{,}57 \times H) - (5{,}07 \times A) - (1{,}76 \times BM) + 309m \quad (1.22)$$

where: ref6MWD—reference value for six-minute walk distance (m); H—height (cm); BM—body mass (kg); A—age (years).

The distance covered during the 6MWT can be translated into exercise prescription with 80% of the average 6MWD.

Furthermore, the final report can attain additional specific indices, e.g., average six-minute walk speed (6MWS in m $\times$ $s^{-1}$) calculated as shown below:

$$6MWSpeed = \frac{6MWD}{360} \quad (1.23)$$

where: 6MWSpeed, average six-minute walk speed ($m \times s^{-1}$); 6MWD, six-minute walk distance (m), 360, 6 min = 6 × 60 s.

Other possible components of the final report include six-minute walk work (6MWW kg × m), heart rate reserve (HRR), heart rate recovery (HRrecov), or estimated peak oxygen uptake (eVO2) and peak work rate (ePWR). The last two estimations have been based on the following formulas [56]:

$$eVO_{2peak} = 4.948 + 0.023 \times 6MWD$$

where: $eVO_{2\text{peak}}$, estimated peak oxygen uptake ($mL \times kg^{-1} \times min^{-1}$); 6MWD, six-minute walk distance (m).

$$ePWR = 0.122 \times 6MWD + 0.72683 \times H - 117.109$$

where: ePWR, estimated peak work rate (W); 6MWD, for six-minute walk distance (m); H, height (cm).

The following values correlated with poor prognostic in ECLIPSE study [57]: age of 65 or more, BMI of 23 or less, 6MWD < 334 m, 6MWSpeed below 0.9 $m \times s^{-1}$, and 6MWW < 20 000.

### 1.3.2 The Two-Minute Step Test

The two-minute step test (2 MST) is a simple alternative for the 6MWT. It serves as a component of the Fullerton Test (Senior Fitness Test) in case of unavailability of walking area or such area is trafficked.

The test involves marching in place as fast as possible for 2 min with the knees lifted between the patella and iliac crest level. Unlike the 6MWT, patients can slow down or rest [58].

Contraindications and reasons for test termination are like 6MWT. The described 2MST norm is 65 steps or more.

### 1.3.3 The Incremental Shuttle Walk Test (ISWT)

The Incremental Shuttle Walking Test is a walking test in which the initial walking pace is gradually increased at regular intervals following a voice signal. The test is discontinued once the patient is too tired or cannot maintain the required speed.

Sufficient information must be delivered to the patient before the test. The contraindications and reasons for test termination are as mentioned above.

The test outcome is the number of completed shuttles (total distance covered). A correlation between the number of shuttles completed and the maximum oxygen uptake has been demonstrated, suggesting the ISWT is a tool for assessing changes in patients' functional capacity during cardiac rehabilitation. The ISWT distance covered by men and women of 395 and 269 m, respectively, has been documented [59]. The ISWT results can be incorporated into exercise training prescription, with an intensity equal to 70% of the peak ISWT speed [60].

**Competing Interests** The authors have no conflicts of interest to declare that are relevant to the content of this chapter.

## References

1. Wasserman K. Diagnosing cardiovascular and lung pathophysiology from exercise gas exchange. Chest. 1997;112:1091–101. https://doi.org/10.1378/chest.112.4.1091.
2. Kaminsky LA, Arena R, Myers J, Peterman JE, Bonikowske AR, Harber MP, et al. Updated reference standards for cardiorespiratory fitness measured with cardiopulmonary exercise testing: data from the fitness registry and the importance of exercise national database (FRIEND). Mayo Clin Proc. 2022;97:285–93. https://doi.org/10.1016/j.mayocp.2021.08.020.
3. Glass S, Dwyer G. ACSM's metabolic calculations handbook. Lippincott Williams & Wilkins 2007; p 22
4. Myers J, Buchanan N, Walsh D, Kraemer M, McAuley P, Hamilton-Wessler M, Froeli-cher VF. Comparison of the ramp versus standard exercise protocols. J Am Coll Cardiol. 1991;17:1334–42. https://doi.org/10.1016/s0735-1097(10)80144-5.
5. Kokkinos P, Kaminsky LA, Arena R, Zhang J, Myers J. New generalized equation for predicting maximal oxygen uptake (from the fitness registry and the importance of exercise national Database). Am J Cardiol. 2017;120:688–92. https://doi.org/10.1016/j.amjcard.2017.05.037.
6. Jang WY, Kang DO, Park Y, Lee J, Kim W, Choi JY, et al. Validation of FRIEND and ACSM equations for cardiorespiratory fitness: comparison to direct measurement in CAD patients. J Clin Med. 2020;9:1889. https://doi.org/10.3390/jcm9061889.PMID:32560313.
7. Kokkinos P, Kaminsky LA, Arena R, Zhang J, Myers J. A new generalized cycle ergometry equation for predicting maximal oxygen uptake: the fitness registry and the importance of exercise national database (FRIEND). Eur J Prev Cardiol. 2018;25:1077–82. https://doi.org/10.1177/2047487318772667.
8. Jetté M, Sidney K, Blümchen G. Metabolic equivalents (METS) in exercise testing, exercise prescription, and evaluation of functional capacity. Clin Cardiol. 1990;13:555–65. https://doi.org/10.1002/clc.4960130809.
9. Gibbons RJ, Balady GJ, Bricker JT, Chaitman BR, Fletcher GF, Froelicher VF, et al. American college of cardiology/American heart association task force on practice guidelines. Committee to update the 1997 exercise testing guidelines. ACC/AHA 2002 guideline update for exercise testing: summary article. A report of the American College of Cardiology/American Heart Association Task Force on Practice Guidelines (Committee to Update the 1997 Exercise Testing Guidelines). J Am Coll Cardiol. 2002;40:1531–40. https://doi.org/10.1016/s0735-1097(02)02164-2.
10. Fletcher GF, Ades PA, Kligfield P, Arena R, Balady GJ, Bittner VA, et al. American heart association exercise, cardiac rehabilitation, and prevention committee of the council on clinical cardiology, council on nutrition, physical activity and metabolism, council on cardiovascular and stroke nursing, and council on epidemiology and prevention. Exercise standards for testing and training: a scientific statement from the American heart association. Circulation. 2013;128:873–934. https://doi.org/10.1161/CIR.0b013e31829b5b44.
11. Myers J, Arena R, Franklin B, Pina I, Kraus WE, McInnis K, Balady GJ. American heart association committee on exercise, cardiac rehabilitation, and prevention of the council on clinical

cardiology, the council on nutrition, physical activity, and metabolism, and the council on cardiovascular nursing. recommendations for clinical exercise laboratories: a scientific statement from the American heart association. Circulation. 2009;119:3144–61. https://doi.org/10.1161/CIRCULATIONAHA.109.192520.
12. Myers J, Do D, Herbert W, Ribisl P, Froelicher VF. A nomogram to predict exercise capacity from a specific activity questionnaire and clinical data. Am J Cardiol. 1994;73:591–6. https://doi.org/10.1016/0002-9149(94)90340-9.
13. Jurca R, Jackson AS, LaMonte MJ, Morrow JR Jr, Blair SN, Wareham NJ, et al. Assessing cardiorespiratory fitness without performing exercise testing. Am J Prev Med. 2005;29:185–93. https://doi.org/10.1016/j.amepre.2005.06.004.
14. Peterman JE, Arena R, Myers J, Ades PA, Bonikowske AR, Harber MP, et al. A Non-exercise prediction of peak oxygen uptake for patients with cardiovascular disease: data from the fitness registry and the importance of exercise international database (Friend). J Cardiopulm Rehabil Prev. 2023;43:115–21. https://doi.org/10.1097/HCR.0000000000000722.
15. Morris CK, Myers J, Froelicher VF, Kawaguchi T, Ueshima K, Hideg A. Nomogram based on metabolic equivalents and age for assessing aerobic exercise capacity in men. J Am Coll Cardiol. 1993;22:175–82. https://doi.org/10.1016/0735-1097(93)90832-L.
16. Gulati M, Black HR, Shaw LJ, Arnsdorf MF, Merz CN, Lauer MS, et al. The prognostic value of a nomogram for exercise capacity in women. N Engl J Med. 2005;353:468–75. https://doi.org/10.1056/NEJMoa044154.
17. Pescatello LS, Arena R, Riebe D, Thompson PD. ACSM's guidelines for exercise testing and prescription. 9th ed. Wolters Kluwer/Lippincott Williams & Wilkins, Philadelphia, PA. 2014, p 84
18. Weber KT, Kinasewitz GT, West JS, Janicki JS, Reichek N, Fishman AP. Long-term vasodilator therapy with trimazosin in chronic cardiac failure. N Engl J Med. 1980;303:242–50. https://doi.org/10.1056/NEJM198007313030502.
19. McInnis KJ, Balady GJ. Comparison of submaximal exercise responses using the Bruce vs modified Bruce protocols. Med Sci Sports Exerc. 1994:103–7. PMID: 8133729.
20. Okin PM, Ameisen O, Kligfield P. A modified treadmill exercise protocol for computer-assisted analysis of the ST segment/heart rate slope: methods and reproducibility. J Electrocardiol. 1986;19:311–8. https://doi.org/10.1016/s0022-0736(86)81058-5.
21. Lehmann G, Schmid S, Ammer R, Schömig A, Alt E. Evaluation of a new treadmill exercise protocol. Chest. 1997;112:98–106. https://doi.org/10.1378/chest.112.1.98.
22. Tamesis B, Stelken A, Byers S, Shaw L, Younis L, Miller DD, Chaitman BR. Comparison of the asymptomatic cardiac ischemia pilot and modified asymptomatic cardiac ischemia pilot versus Bruce and Cornell exercise protocols. Am J Cardiol. 1993;72:715–20. https://doi.org/10.1016/0002-9149(93)90891-f.
23. Ellestad MH, Allen W, Wan MC, Kemp GL. Maximal treadmill stress testing for cardiovascular evaluation. Circulation. 1969;39:517–22. https://doi.org/10.1161/01.cir.39.4.517.
24. Bruce RA, Blackmon JR, Jones JW, Strait G. Exercising testing in adult normal subjects and cardiac patients. Pediatrics. 1963;32(SUPPL):742–56 PMID: 14070531.
25. Myers J, Buchanan N, Smith D, Neutel J, Bowes E, Walsh D, Froelicher VF. Individualized ramp treadmill. Observations on a new protocol Chest. 1992;101(5 Suppl):236S-241S PMID: 1576842.
26. Kaminsky LA, Whaley MH. Evaluation of a new standardized ramp protocol: the BSU/Bruce Ramp protocol. J Cardiopulm Rehabil. 1998;18:438–44. https://doi.org/10.1097/00008483-199811000-00006.
27. American Association of Cardiovascular and Pulmonary Rehabilitation. Guidelines for cardiac rehabilitation and secondary prevention programs. 4 ed. Human Kinetics. 2004, p 79
28. Pescatello LS, Arena R, Riebe D, Thompson PD. ACSM's guidelines for exercise testing and prescription. 9th ed. Wolters Kluwer/Lippincott Williams & Wilkins, Philadelphia, PA. 2014, p 124
29. Arena R, Myers J, Williams MA, Gulati M, Kligfield P, Balady GJ, Collins E, Fletcher G; American heart association committee on exercise, rehabilitation, and prevention of the council on

clinical cardiology; American heart association council on cardiovascular nursing. Assessment of functional capacity in clinical and research settings: a scientific statement from the American heart association committee on exercise, rehabilitation, and prevention of the council on clinical cardiology and the council on cardiovascular nursing. Circulation. 2007;116:329–43. https://doi.org/10.1161/CIRCULATIONAHA.106.184461.
30. Rybicki J. Zasady doboru protokołu testu wysiłkowego u chorych i zdrowych osób nietrenujących wyczynowo. In: Pikto-Pietkiewicz W, Piotrowicz R, Dłużniewski M, Kalarus Z, Średniawa B, editors. Sport wyczynowy i rekreacyjny—problemy kardiologa i interni-sty. 2nd ed. Czelej; 2017. pp 51–69
31. Beltz NM, Gibson AL, Janot JM, Kravitz L, Mermier CM, Dalleck LC. Graded exercise testing protocols for the determination of VO2max: historical perspectives, progress, and future considerations. J Sports Med (Hindawi Publ Corp). 2016;2016:3968393. https://doi.org/10.1155/2016/3968393.
32. American Thoracic Society; American College of Chest Physicians. ATS/ACCP Statement on cardiopulmonary exercise testing. Am J Respir Crit Care Med. 2003;167:211–77. https://doi.org/10.1164/rccm.167.2.211
33. Mezzani A. Cardiopulmonary exercise testing: basics of methodology and measurements. Ann Am Thorac Soc. 2017;14(Supplement_1):S3–S11. https://doi.org/10.1513/AnnalsATS.201612-/997FR
34. Balady GJ, Arena R, Sietsema K, Myers J, Coke L, Fletcher GF, et al. American heart association exercise, cardiac rehabilitation, and prevention committee of the council on clinical cardiology; council on epidemiology and prevention; council on peripheral vascular disease; interdisciplinary council on quality of care and outcomes research. Clinician's Guide to cardiopulmonary exercise testing in adults: a scientific statement from the American Heart Association. Circulation. 2010;122:191–225. https://doi.org/10.1161/CIR.0b013e3181e52e6937.
35. Campbell SC. A comparison of the maximum voluntary ventilation with the forced expiratory volume in one second: an assessment of subject cooperation. J Occup Med. 1982;24:531–3 PMID: 7119913.
36. Hansen JE, Casaburi R, Cooper DM, Wasserman K. Oxygen uptake as related to work rate increment during cycle ergometer exercise. Eur J Appl Physiol Occup Physiol. 1988;57:140–5. https://doi.org/10.1007/BF00640653.
37. Howley ET, Bassett DR Jr, Welch HG. Criteria for maximal oxygen uptake: review and commentary. Med Sci Sports Exerc. 1995;27:1292–301 PMID: 8531628.
38. Mezzani A, Agostoni P, Cohen-Solal A, Corrà U, Jegier A, Kouidi E, et al. Standards for the use of cardiopulmonary exercise testing for the functional evaluation of cardiac patients: a report from the exercise physiology section of the European association for cardiovascular prevention and rehabilitation. Eur J Cardiovasc Prev Rehabil. 2009;16:249–67. https://doi.org/10.1097/HJR.0b013e32832914c8.
39. Wasserman K, Hansen JE, Sue DY, Stringer WW, Whipp BJ. Normal values. In: Was-serman K, Hansen JE, Sue DY, Stringer WW, Whipp BJ, editors. Principles of exercise testing and interpretation: including pathophysiology and clinical applications. Philadelphia: Lippincott Williams & Wilkins; 2005. p. 160–82.
40. de Souza E, Silva CG, Kaminsky LA, Arena R, Christle JW, Araújo CGS, Lima RM, et al. A reference equation for maximal aerobic power for treadmill and cycle ergometer exercise testing: analysis from the FRIEND registry. Eur J Prev Cardiol. 2018;25:742–50. https://doi.org/10.1177/2047487318763958.
41. Gläser S, Koch B, Ittermann T, Schäper C, Dörr M, Felix SB, et al. Influence of age, sex, body size, smoking, and beta blockade on key gas exchange exercise parameters in an adult population. Eur J Cardiovasc Prev Rehabil. 2010;17:469–76. https://doi.org/10.1097/HJR.0b013e328336a124.
42. Takken T, Mylius CF, Paap D, Broeders W, Hulzebos HJ, Van Brussel M, Bongers BC. Reference values for cardiopulmonary exercise testing in healthy subjects—an updated systematic review. Expert Rev Cardiovasc Ther. 2019;17:413–26. https://doi.org/10.1080/14779072.2019.1627874. Epub 2019 Jun 11.

43. Guazzi M, Adams V, Conraads V, Halle M, Mezzani A, Vanhees L, Arena R, Fletcher GF, Forman DE, Kitzman DW, Lavie CJ, Myers J. EACPR; AHA. EACPR/AHA joint scientific statement. Clinical recommendations for cardiopulmonary exercise testing data assessment in specific patient populations. Eur Heart J. 2012;33:2917–27. https://doi.org/10.1093/eurheartj/ehs221. Epub 2012 Sep 5.
44. Baba R, Nagashima M, Goto M, Nagano Y, Yokota M, Tauchi N, Nishibata K. Oxygen uptake efficiency slope: a new index of cardiorespiratory functional reserve derived from the relation between oxygen uptake and minute ventilation during incremental exercise. J Am Coll Cardiol. 1996;28:1567–72. https://doi.org/10.1016/s0735-1097(96)00412-3.
45. Myers J, Arena R, Dewey F, Bensimhon D, Abella J, Hsu L, et al. A cardiopulmonary exercise testing score for predicting outcomes in patients with heart failure. Am Heart J. 2008;156:1177–83. https://doi.org/10.1016/j.ahj.2008.07.010.
46. Arena R, Myers J, Abella J, Peberdy MA, Bensimhon D, Chase P, Guazzi M. Development of a ventilatory classification system in patients with heart failure. Circulation. 2007;115:2410–7. https://doi.org/10.1161/CIRCULATIONAHA.107.686576.
47. Wasserman K, Hansen JE, Sue DY, Whipp BJ, Casaburi R. Principles of exercise testing and interpretation: including pathophysiology and clinical applications. 5th ed. Philadelphia: Lippincott Williams & Wilkins; 2012.
48. Dumitrescu D, Rosenkranz S. Graphical data display for clinical cardiopulmonary exercise testing. Ann Am Thorac Soc. 2017;14(Supplement_1):S12–S21. https://doi.org/10.1513/AnnalsATS.201612-955FR.
49. Hansen D, Bonné K, Alders T, Hermans A, Copermans K, Swinnen H, et al. Exercise training intensity determination in cardiovascular rehabilitation: should the guidelines be reconsidered? Eur J Prev Cardiol. 2019;26:1921–8. https://doi.org/10.1177/2047487319859450.
50. Anselmi F, Cavigli L, Pagliaro A, Valente S, Valentini F, Cameli M, et al. The importance of ventilatory thresholds to define aerobic exercise intensity in cardiac patients and healthy subjects. Scand J Med Sci Sports. 2021;31:1796–808. https://doi.org/10.1111/sms.14007.
51. Hansen D, Abreu A, Ambrosetti M, Cornelissen V, Gevaert A, Kemps H, et al. Exercise intensity assessment and prescription in cardiovascular rehabilitation and beyond: why and how: a position statement from the secondary prevention and rehabilitation section of the European association of preventive cardiology. Eur J Prev Cardiol. 2022;29:230–45. https://doi.org/10.1093/eurjpc/zwab007.
52. Guazzi M, Arena R, Halle M, Piepoli MF, Myers J, Lavie CJ. 2016 focused update: clinical recommendations for cardiopulmonary exercise testing data assessment in specific patient populations. Eur Heart J. 2018;39:1144–61. https://doi.org/10.1093/eurheartj/ehw180.
53. Ruzzolini M, Ambrosetti M. Cardiopulmonary exercise testing in cardiac rehabilitation: from the reporting form to structured exercise prescription. A proposal from the Italian alliance for cardiovascular rehabilitation and prevention (Itacare-P). Int J Cardiol Cardiovasc Risk Prev. 2023;18:200191. https://doi.org/10.1016/j.ijcrp.2023.200191
54. Pepera GK, Sandercock GR, Sloan R, Cleland JJ, Ingle L, Clark AL. Influence of step length on 6-minute walk test performance in patients with chronic heart failure. Physiotherapy. 2012;98:325–9. https://doi.org/10.1016/j.physio.2011.08.005.
55. Enright PL, McBurnie MA, Bittner V, Tracy RP, McNamara R, Arnold A et al. Cardiovascular health study. The 6-min walk test: a quick measure of functional status in elderly adults. Chest. 2003;123:387–98. https://doi.org/10.1378/chest.123.2.387
56. Ross RM, Murthy JN, Wollak ID, Jackson AS. The six minute walk test accurately estimates mean peak oxygen uptake. BMC Pulm Med. 2010;10:31. https://doi.org/10.1186/1471-2466-10-31.
57. Andrianopoulos V, Wouters EF, Pinto-Plata VM, Vanfleteren LE, Bakke PS, Franssen FM, Agusti A, MacNee W, Rennard SI, Tal-Singer R, Vogiatzis I, Vestbo J, et al. Prognostic value of variables derived from the six-minute walk test in patients with COPD: results from the ECLIPSE study. Respir Med. 2015;109:1138–46. https://doi.org/10.1016/j.rmed.2015.06.013.
58. Paker N, Şirin Ahısha B, Kesiktaş N, Buğdaycı ND, Soluk ÖY. Construct validity and reliability of the 2-minute step test in patients with lumbar spinal stenosis. Disabil Rehabil. 2024;19:1–6. https://doi.org/10.1080/09638288.2024.2428824.

59. Cardoso FM, Almodhy M, Pepera G, Stasinopoulos DM, Sandercock GR. Reference values for the incremental shuttle walk test in patients with cardiovascular disease entering exercise-based cardiac rehabilitation. J Sports Sci. 2017;35:1–6. https://doi.org/10.1080/02640414.2016.1151925.
60. Zainuldin R, Mackey MG, Alison JA. Prescription of walking exercise intensity from the 6-minute walk test in people with chronic obstructive pulmonary disease. J Cardiopulm Rehabil Prev. 2015;35:65–9. https://doi.org/10.1097/HCR.0000000000000074.

# Chapter 2
# Exercise Programming

**Jerzy Rybicki and Adam Staron**

**Abstract** An initial assessment and process of risk stratification should precede the commencement of exercise training prescription. An early assessment allows for the identification of the individual needs of patients who are referred to cardiac rehabilitation and development of the personalized plan of care. In contrast, cardiac risk stratification seeks to identify patients who are at risk of experiencing a recurrence of a cardiac event.

**Keywords** Exercise intensity · Oxygen uptake · Heart rate reserve · Talk test · Borg scale

## 2.1 Risk Stratification

The group of patients referred to cardiac rehabilitation (CR) is significantly heterogeneous; due to the main diagnosis, the following groups of patients can be distinguished:

- post-myocardial infarction (MI)/post-acute coronary syndrome (ACS) treated percutaneously and/or surgically,
- after valve repair or replacement,
- after implanted cardiac electrical devices (ICED) (permanent pacemaker (PPM), cardiac resynchronization therapy device (CRT), and implantable cardioverter-defibrillator (ICD),
- after heart and/or lung transplantation,
- with chronic heart failure (CHF).

J. Rybicki
REPTY" Upper Silesian Rehabilitation Center Tarnowskie Góry Poland, Silesian Center for Heart Disease, Zabrze, Poland
e-mail: j_rybicki@cyberia.pl

A. Staron (✉)
The Upper Silesian Medical Center, Katowice, Poland
e-mail: astaron1973@wp.pl



J. Rybicki and A. Staron (eds.), *Exercise Programming for Cardiac Patients*,
https://doi.org/10.1007/978-3-031-88475-7_2

Elderly patients, often belonging to several of the above-mentioned indication groups, with severe comorbidities (diabetes, peripheral arterial disease) and numerous risk factors for cardiovascular diseases (dyslipidemia, hypertension, obesity) are currently a common challenge for rehabilitation professionals.

The admission assessment embraces a clinical evaluation and tests. The clinical evaluation should incorporate the review of the event course, the symptoms declared, the presence of cardiovascular risk factors, comorbidities, and the medical treatment.

In case of implanted cardiac electrical devices, the device characteristics, intervention mode, and thresholds should be studied.

Psychological screening should be performed adopting a questionnaire or a scale. The physical examination should be performed according to the main diagnoses of the patient. At entry the resting 12-lead electrocardiogram (ECG), routine laboratory testing, resting transthoracic echocardiography (ECHO), 24 h ambulatory ECG monitoring (Holter), and functional capacity testing should be conducted. Screening transthoracic ECHO on entry allows for exclusion of pericardial fluid, or intracardiac thrombus. The ECHO study is recommended at the end of phase II of cardiac rehabilitation (CR) to evaluate left ventricular systolic and diastolic function, particularly for determining the need for cardioverter-defibrillator implantation [1].

The measurement of cardiorespiratory fitness (CRF) should be integrated into clinical practice. Meta-analysis representing over 20.9 million observations from 199 cohort studies demonstrated that the increase of CRF by 1 MET resulted in a reduction in the risk of all-cause, cardiovascular, and cancer mortality [2].

Exercise-related acute cardiac events generally occur in people with already existing structural cardiac abnormality. In young individuals, the most frequent findings are hypertrophic cardiomyopathy and coronary artery anomaly, whereas coronary artery disease dominates in older individuals who died during exercise [3]. An acute cardiovascular event occurs extremely rarely during vigorous exertion (1 in 500,000 to 1 in 2,600,000 h of exercise [4].

An exercise dose–response relationship between exercise and mortality includes the most significant benefits when physically inactive individuals become more physically active. However, extreme exercise can provoke atrial and ventricular arrhythmias, pathological remodeling of the myocardium, and coronary artery calcification. The risk-protection paradox has been demonstrated in numerous studies with a vigorous intensity that increases the short-term risk for an adverse cardiac event, e.g., myocardial infarction, while significantly reducing the long-term risk for premature mortality [5]. The beneficial effects of regular exercise undeniably outweigh the risks. Thus, exercise-related risk and adverse events stratification is fundamental to assess the risk of future cardiac events. The risk stratification formula commonly used is developed by the American Association of Cardiovascular and Pulmonary Rehabilitation [6].

Similarly, many national and international cardiology societies have developed risk stratification protocols for exercise-induced cardiac events. The European Society of Cardiology (ESC) recommends different left ventricular ejection fraction (LVEF) thresholds (<35%, 35–49%, and 50% or more) for high, moderate, and

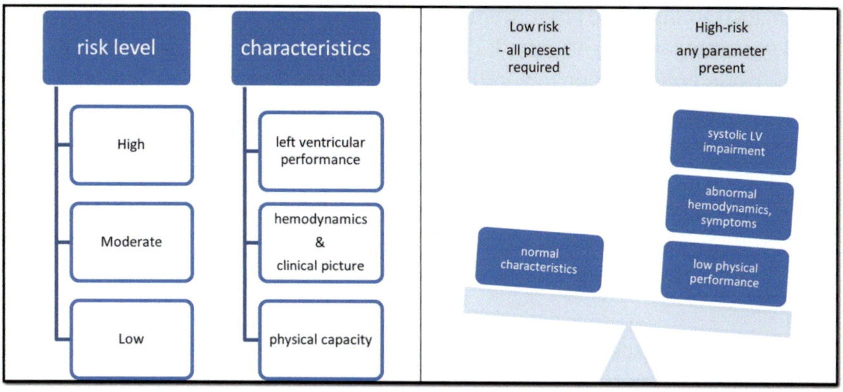


**Fig. 2.1** Risk stratification for cardiac events during exercise: levels, characteristics and evaluation, by the authors LV, left ventricle

low risk, respectively [7]. The physical capacity thresholds for low/moderate and moderate/high-risk are 7 MET and 5 MET, respectively.

The general principles of the risk stratification protocol for cardiac incidents during training are presented in Fig. 2.1.

### *2.1.1 Principles of Training Programming*

The exercise training program should incorporate the following:

- aerobic endurance training (walking, walking with sticks, biking, training on a stepper, running, swimming)
- dynamic resistance training (on multi-weight machines or with free weights, dumbbells, and elastic bands)
- stretching exercise
- balance and coordination exercises
- relaxation session

The exercise training session should commence with a warm-up, followed by the conditioning phase, and finish with a cool-down component:

- the warm-up phase typically lasts for 3–10 min, including very light-to-light intensity exercise at 20%–40% of heart rate reserve, <11 on the RPE Borg scale. The warm-up allows for gradual body adaptation to the exertion and precludes the sudden rise of catecholamines. It is suggested that by the end of the warm-up phase, an intensity of 40% of the heart rate reserve (or 10 on the RPE scale) should be reached. The commonly performed warm-up activities are marching on the spot, walking, or low-intensity cycling

- the main training phase (conditioning) is performed in continuous or interval mode and typically lasts 20–40 min. It can be performed using one specific equipment (e.g., cycle ergometer or treadmill) or can take the form of circuit training with dedicated stations
- the cool-down phase lasts 3–10 min, involves exercises at low-to-moderate-intensity. The cool-down phase allows for the gradual recovery of heart rate and blood pressure; thus, precludes post-exertional events such as ischemia, arrhythmia, or hypotension, which can occur within few min of exercise cessation. All patients should be supervised for a minimum of 15 min after the cool-down phase.

Exercise training prescription should be consistent with the FITT-VP principle, which describes [1, 8]:

- frequency (F)—the number of training sessions per time unit, e.g., per week
- intensity (I)—objective and subjective methods of exercise load
- time (T)—duration of the session and its phases
- type (T)—mode and protocol of training
- volume (V)—total energy expenditure during training and per week
- progression (P)—method and rate of training load increase.

Frequency—physical activity is recommended on most days of the week.

Intensity—should be preferably determined based on the cardiopulmonary exercise testing in relation to ventilatory thresholds. Due to the limited availability of cardiopulmonary analysis, alternative methods of exercise intensity determination are utilized, which are based on the rating of perceived exertion (Borg scale), training heart rate (heart rate reserve or % of the maximal heart rate), MET reserve, or peak work rate.

Time—the goal for aerobic training duration is 30–60 min, whereas strength, flexibility, and neuromotor training sessions commonly last 15–20 min.

Type—aerobic, resistance, stretching, neuromotor and relaxation exercises.

Volume—the product of frequency, intensity, and time (frequency × intensity × time).

Progression—is based on clinical status, fitness level, and training adaptation. The training duration is usually extended by 5–10 min over 1–2 weeks. Typically, the training duration increase precedes an increase in intensity or frequency (Fig. 2.2).

FITT-based training principles are strictly related to training safety, as shown in Fig. 2.3.

### 2.1.2 Aerobic Training Intensity Programming

Determination of training intensity can be executed based on the results of the cardiopulmonary exercise test (CPET) (optimal standards) or electrocardiographic exercise test (ExT) (minimal standards) [9] and a carefully conducted analysis of the risk

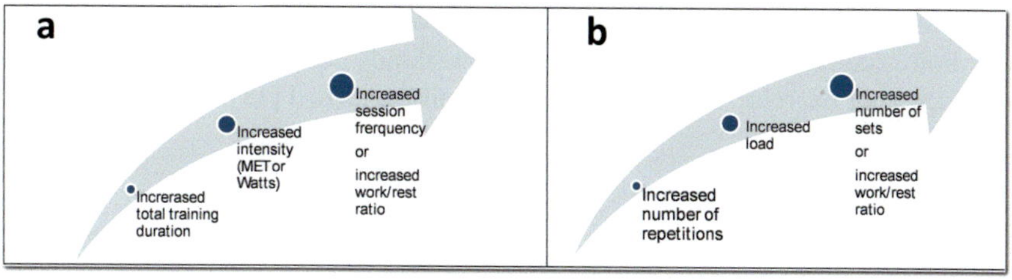


**Fig. 2.2** **a** typical progression of aerobic and **b** resistance training, by the authors

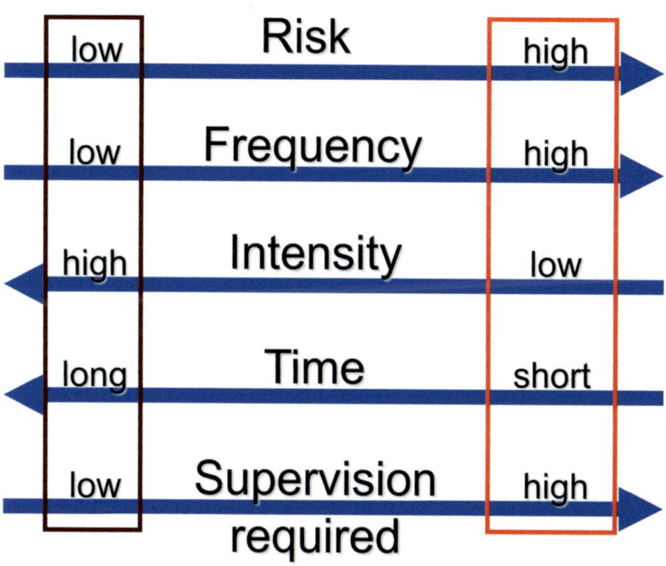


**Fig. 2.3** Correlation of risk and training parameters and the requirement for its monitoring, by the authors

of cardiac events during exercise, using the objective indirect, direct and subjective indices. Field tests should be considered complementary methods of functional capacity assessment.

Generally, there are two ways of training intensity programming listed below:

- "Threshold-based" approach, based on the determined metabolic ventilation thresholds VT1 and VT2, allows for the distinction of three [9, 10] or four [11] training intensity zones,
- the "range-based" method expressed as a percentage of the peak tolerated load (watt), measured oxygen uptake (VO2) or estimated metabolic equivalent of task (MET), and maximal heart rate (HRmax) [8, 12]

According to the position statement of the European and American cardiac rehabilitation experts, based on the ventilation thresholds determined in the CPET, the following four exercise intensity domains are distinguished [11]:

- light to moderate,
- moderate to high,
- high to severe,
- severe to extreme

Light to moderate intensity aerobic exercises include work rates (WR) with energy demand below the first ventilation threshold (VT1). The duration of continuous training may be >30 min.

The moderate to high-intensity domain includes efforts between the VT1 and critical power (CP), corresponding to the highest workload at which exercise can be continued under conditions of constant VO2 and blood lactate concentration. Due to the complexity of determining CP (i.e. the need to perform few tests with constant load), it is assumed that its value equals 60%-70% of peak work rate (PWR). The duration of continuous training at this intensity can be programmed for 20–30 min.

The high to severe range comprises workloads from CP to VO2peak without reaching a maximal steady state. This intensity can be used in interval training with load cycles lasting 3–20 min.

The intensity domain from severe to extreme includes efforts at the level of maximum steady state VO2, leading to effort termination in less than 3 min. This intensity is used for the short and moderate duration of high-intensity interval training load cycles.

From a practical point of view, personalized exercise training intensities, expressed as METs, representing optimal fitness levels and associated with good health outcomes and increased long-term survival, should be established. To deliver this, the age- and sex-based regression equation, utilizing the Fitness Registry and the Importance of Exercise: A National Database (FRIEND), correlating with 70% oxygen uptake reserve, has been developed [13].

The ACSM [8] and European Society of Cardiology (ESC) [12] exercise training intensity "range-based" thresholds have been demonstrated in Tables 2.1, 2.2, 2.3, and 2.4.

Determination of training intensity can be executed by indicators listed in Fig. 2.4.

### *2.1.3 Training Heart Rate Calculation Methods*

The observed straight-linear correlation of VO2 with HR in the normal chronotropic response to submaximal exercise (Fig. 2.5) is the basis for using training heart rate (THR) as an indirect index of training intensity.

This frequently used parameter, due to the ease of training monitoring, is not applicable or is limited in many clinical cases, such as:

**Table 2.1** Classification of exercise intensity: relative exercise intensity for cardiorespiratory endurance exercise according to the American College of Sports Medicine (with permission from Ref. [8])

| Intensity | % HRR or %$VO_2R$ | %$HR_{max}$ | %$VO_{2max}$ | Perceived exertion RPE Scale (6–20) |
|---|---|---|---|---|
| Very light | <30 | <57 | >37 | RPE < 9 |
| Light | 30–39 | 57–63 | 37–45 | RPE 9–11 |
| Moderate | 40–59 | 64–76 | 46–63 | RPE 12–13 |
| Vigorous | 60–89 | 77–95 | 64–90 | RPE 14–17 |
| Near maximal to maximal | ≥90 | ≥96 | ≥91 | RPE ≥ 18 |

%HRR, percent of heart rate reserve; %$VO_2R$, percent of oxygen uptake reserve; %$HR_{max}$, percent of maximal heart rate; %$VO_{2max}$, percent of maximal oxygen uptake; RPE, ratings of perceived exertion $HR_{max}$, maximal heart rate; $VO_{2max}$, maximal oxygen uptake

**Table 2.2** Classification of exercise intensity: Intensity (%$VO_{2max}$) relative to maximal exercise capacity in METs for cardiorespiratory endurance exercise and resistance exercise according to the American College of Sports Medicine (with permission from Ref. [8])

| Intensity | Cardiorespiratory endurance exercise | | | Resistance exercise |
|---|---|---|---|---|
| | 20 METs %$VO_{2max}$ | 10 METs %$VO_{2max}$ | 5 METs %$VO_{2max}$ | % 1RM |
| Very light | <34 | <37 | >44 | <30 |
| Light | 34–42 | 37–45 | 44–51 | 30–49 |
| Moderate | 43–61 | 46–63 | 52–67 | 50–69 |
| Vigorous | 62–90 | 64–90 | 68–91 | 70–84 |
| Near maximal to maximal | ≥91 | ≥91 | ≥92 | ≥85 |

METs, metabolic equivalent of tasks; %$VO_{2max}$, percent of maximal oxygen uptake; % 1RM, percent of one-repetition maximum

**Table 2.3** Classification of exercise intensity: absolute exercise intensity for cardiorespiratory endurance exercise according to the American College of Sports Medicine (with permission from Ref. [8])

| Intensity | Absolute intensity | Absolute Intensity (MET) by Age | | |
|---|---|---|---|---|
| | METs | Young (20–39 yr) | Middle-aged (40–64 yr) | Older (≥65 yr) |
| Very light | <2.0 | <2.4 | <2.0 | <1.6 |
| Light | 2.0–2.9 | 2.4–4.7 | 2.0–3.9 | 1.6–3.1 |
| Moderate | 3.0–5.9 | 4.8–7.1.9 | 4.0–5.9 | 3.2–4.7 |
| Vigorous | 6.0–8.7 | 7.2–10.1 | 6.0–8.4 | 4.8–6.7 |
| Near maximal to maximal | ≥8.8 | ≥10.2 | ≥8.5 | ≥6.8 |

METs, metabolic equivalent of tasks; yr, years

**Table 2.4** Indices of exercise intensity for endurance sports from maximal exercise testing and training zones according to the European Society of Cardiology, with permission from reference 12

| Intensity | $VO_{2max}$ (%) | $HR_{max}$ (%) | HRR (%) | RPE Scale | Training zone |
|---|---|---|---|---|---|
| Low intensity, light exercise | <40 | <55 | <40 | 10–11 | Aerobic |
| Moderate intensity exercise | 40–69 | 55–74 | 40–69 | 12–13 | Aerobic |
| High intensity | 70–85 | 75–90 | 70–85 | 14–16 | Aerobic + lactate |
| Very high intensity exercise | >85 | >90 | >85 | 17–19 | Aerobic + lactate + anaerobic |

Low-intensity exercise is below the aerobic threshold; moderate is above the aerobic threshold but not reaching the anaerobic zone; high intensity is close to the anaerobic zone; and very intense exercise is above the anaerobic threshold

$VO_{2max}$, maximum oxygen consumption; $HR_{max}$, maximum heart rate; HRR, heart rate reserve; RPE, rate of perceived exertion in Borg 6–20 points

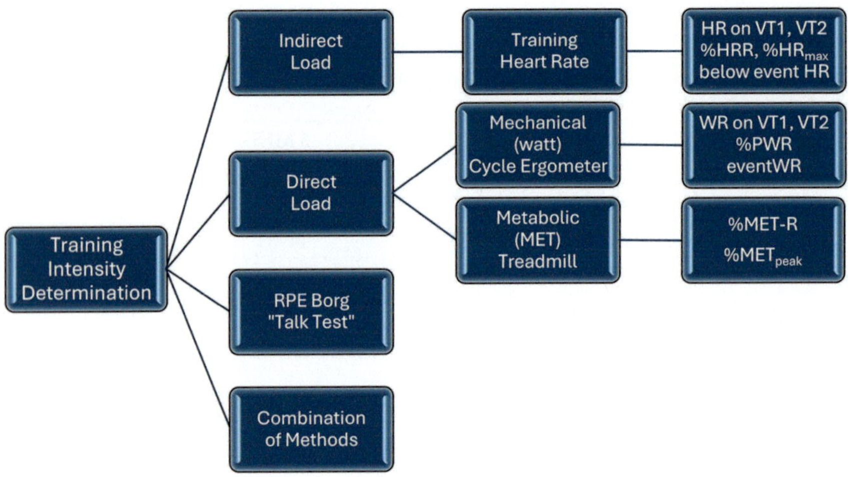


**Fig. 2.4** Methods of training intensity programming, by the authors HR, heart rate; VT1, first ventilatory threshold; VT2, second ventilatory threshold; HRR, heart rate reserve; WR, work rate; PWR, peak work rate; MET, metabolic equivalent of task, MET-R, metabolic equivalent reserve; $MET_{peak}$, peak metabolic equivalent; RPE, rate of perceived exertion

- atrial fibrillation,
- after the implantation of cardiac pacemaker,
- heart transplant recipients,
- chronotropic incompetence.

The following four methods are used to calculate THR:

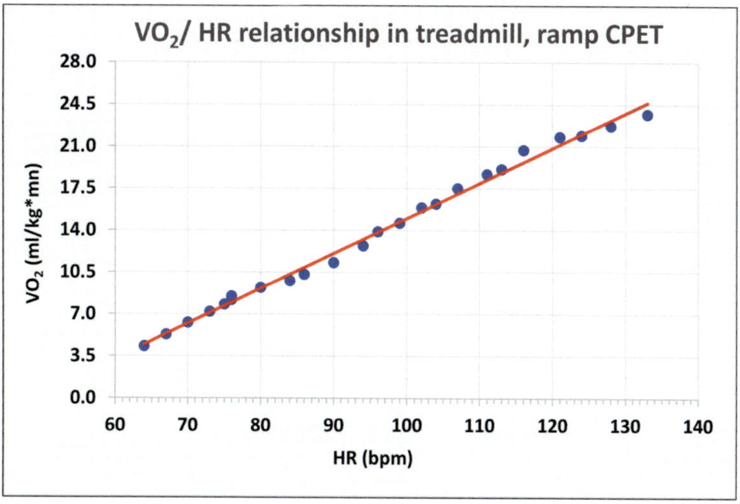


**Fig. 2.5** Oxygen uptake/work rate correlation in treadmill ramp cardiopulmonary exercise test in 50 years men after myocardial infarction, treated with beta-blocker, by the authors $VO_2$, oxygen consumption: HR, heart rate

- HR on VT1 and VT2—heart rate referred to corresponding $VO_2$ on the first (aerobic AeT) and the second (anaerobic AnT) ventilatory threshold determined by CPET. These THR values are used to establish the exercise intensity zones,
- %HRR—percent of heart rate reserve method according to Karvonen [14]:

$$THR = 0.01 \times Int \times (HR_{max} - HR_{rest}) + HR_{rest} \tag{2.1}$$

where: THR, training heart rate (bpm); 0.01, decimal conversion factor for intensity: Int, desired intensity (%); $HR_{max}$, maximal heart rate (bpm); $HR_{rest}$, resting heart rate (bpm).

Percent HRR correlates well with percent $VO_2$ reserve (%$VO_2$R) [15, 16], and their values are used together in the recommended training intensity classes according to ACSM classification (Table 2.1).

- %$HR_{max}$—percent of maximal heart rate attained during symptom-limited exercise test:

$$THR = 0.01 \times Int \times HR_{max} \tag{2.2}$$

where: THR, training heart rate (bpm); 0.01, decimal conversion factor for intensity: Int, desired intensity (%); $HR_{max}$, maximal heart rate (bpm).

Figure 2.6 illustrates THR values calculated using the %HRR, and %$HR_{max}$ methods for the same training intensities, resting and maximal heart rates.

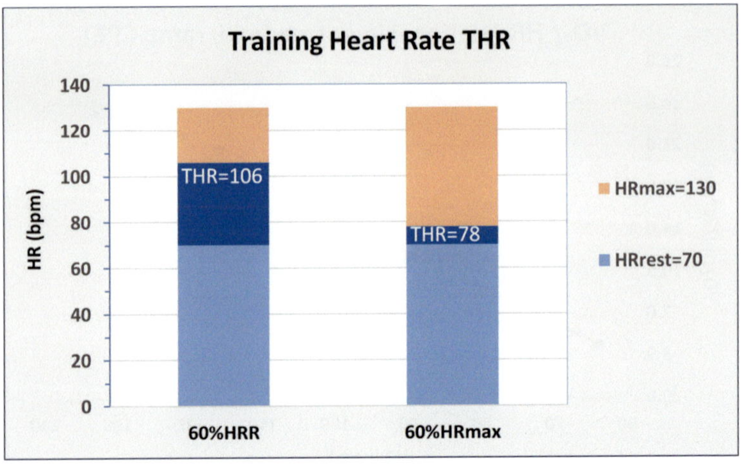


**Fig. 2.6** Two ways to calculate training heart rate, by the authors THR, training heart rate; HRmax, maximal heart rate; HRrest, resting heart rate; %HRR, percent of heart rate reserve method; %HRmax, percent of maximal heart rate method

- Event HR—training heart rate below myocardial ischaemic threshold (MIT) which was defined as the HR or work rate (WR) corresponding to a 1 mm horizontal or down-sloping ST segment depression in ExT [9]:

$$THR(bpm) = eventHR(bpm) - 10(bpm) \tag{2.3}$$

Figure 2.7 shows an example of how to determine event HR in an exercise test.

A more sensitive indicator of MIT is the flattening or decline of the oxygen pulse ($O_2$pulse) trajectory observed in the CPET [17]—Fig. 2.8

Appearance or worsening of cardiac arrhythmias (Fig. 2.9) or conduction abnormalities (Fig. 2.10) should be recorded in the test protocol and taken into account in programming the training intensity.

### 2.1.4 Training Work Rate Calculation Methods

Direct training work rate (TWR) (watt) in cycle training is determined based on a cardiopulmonary or ECG test performed on a cycle ergometer according to the following approaches:

- WR on VT1 and VT2—work rate on aerobic and anaerobic thresholds to determine the load of intensity zones.
- %PWR—percent of peak work rate:

$$TWR = 0.01 \times Int \times PWR \tag{2.4}$$

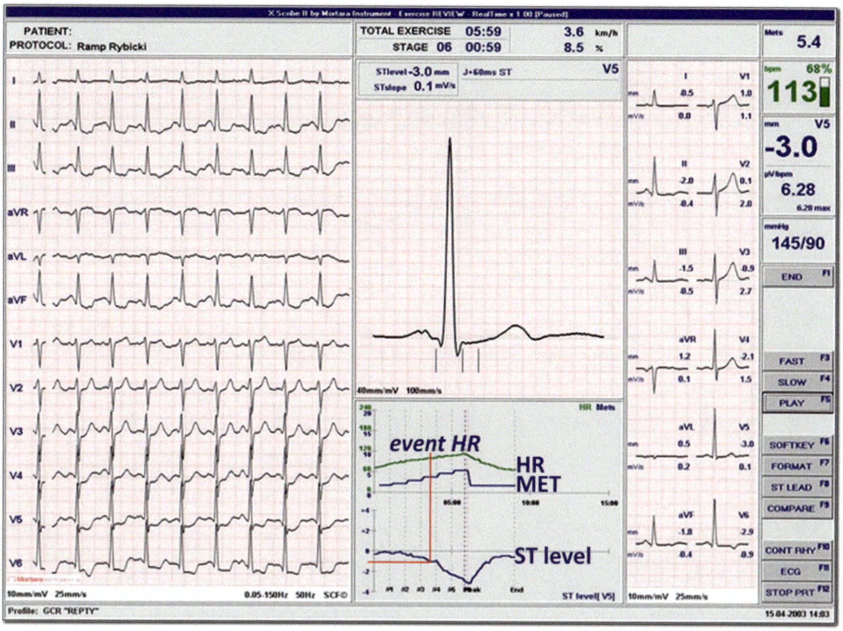


**Fig. 2.7** A practical method for determining the event heart rate on myocardial ischaemic threshold, by the authors

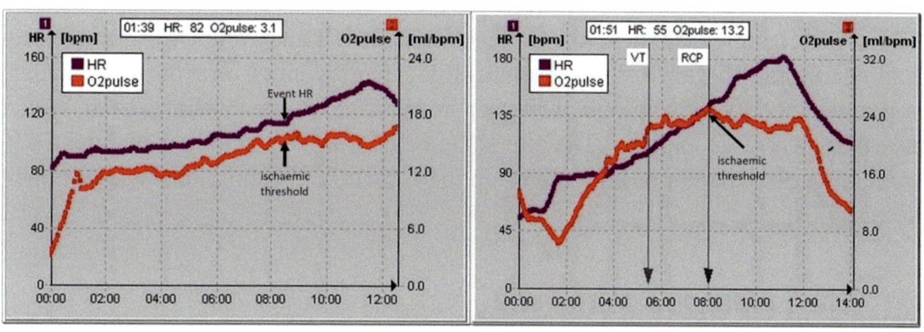


**Fig. 2.8** The beginning of a flattening or decline in the oxygen pulse trajectory indicates (*arrows*) the ischaemic threshold and corresponding event HR during heart rate, by the authors

where: TWR, training work rate (watt); 0.01, decimal conversion factor for intensity: Int, desired intensity (%); PWR, peak work rate (watt).

- Event WR—The watt load associated with the electrocardiographic events previously mentioned for event HR suggests the use of subthreshold loads:

$$WR(watt) = eventWR(watt) - 10(watt) \tag{2.5}$$

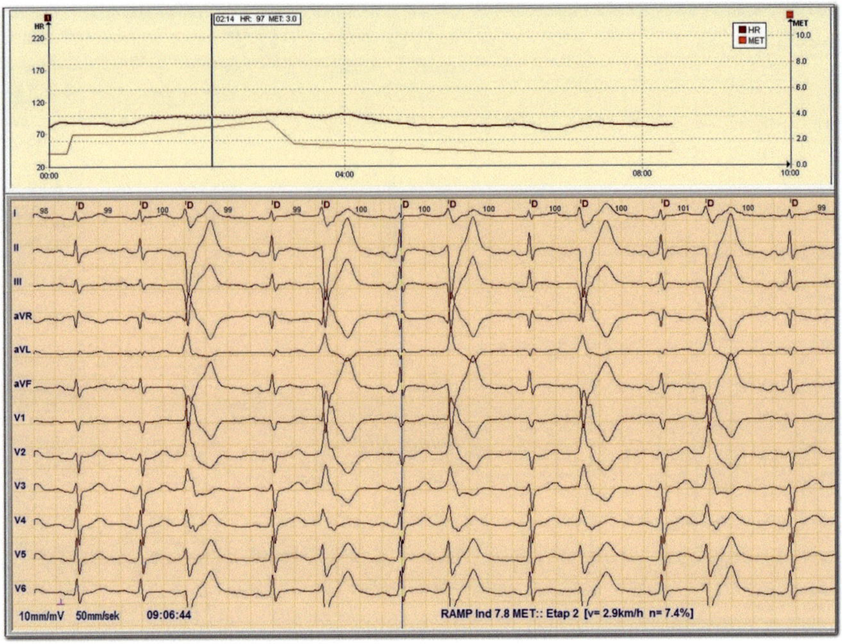

Fig. 2.9 Event threshold (HR = 99 bpm)—appearance of ventricular bigeminy in exercise test, by the authors

However, it should be remembered that the event WR and event HR parameters are "highly unstable" because of the pharmacological intervention after the test and many mechanisms of the physiological impact of exercise on the cardiovascular system known as "warm-up angina." The phenomenon of warm-up angina is manifested by a reduction of ischemia or an increase in the ischemia threshold during subsequent exercise [18].

## Formulas of MET to Watt Calculation

If the exercise test is performed on a treadmill and the training is programmed on a bicycle ergometer, the peak work rate in watts should be estimated (ePWR) based on the maximum exercise capacity expressed in METs. To calculate ePWR value, it is possible to use the transformed formulas for calculating oxygen consumption in the cycle ergometer exercise test according to ACSM [19] (Eq. 2.8) or, with greater accuracy, using the Fitness Registry and the Importance of Exercise National Database (FRIEND) ergometry equation [20] (Eq. 2.11).

In this estimation method, attention should be paid to the method of calculating MET on the treadmill according to ACSM [19] or FRIEND equation [21] and consider the 10–20% higher VO2 in the treadmill compared to the cycle

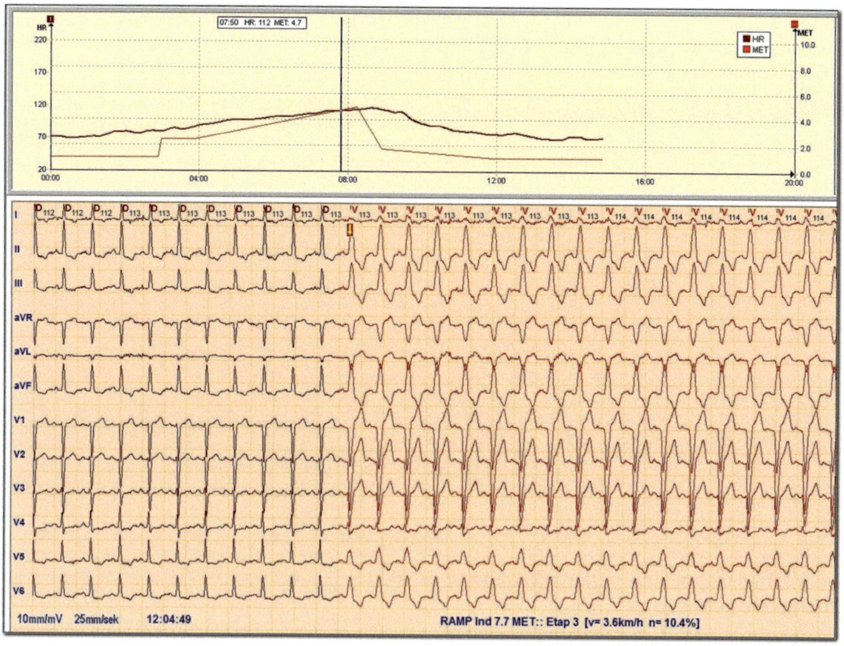


**Fig. 2.10** Event threshold (HR = 113 bpm)—ischemic left bundle branch block in exercise test, by the authors

ergometer test. MET to watt calculation by ACSM ergometry equation [19] (Eq. 2.8) transformation:

$$VO_2 = 11.0 \times \frac{\mathrm{WR}}{\mathrm{BM}} + 7.0$$

$$MET \times 3.5 = 11.0 \times \frac{\mathrm{WR}}{\mathrm{BM}} + 7.0$$

$$MET \times 3.5 - 7.0 = 11.0 \times \frac{WR}{BM}$$

$$BM \times (MET \times 3.5 - 7.0) = 11.0 \times WR$$

$$WR = BM \times (MET \times 3.5 - 7.0)/11{,}0 \tag{2.6}$$

where: $VO_2$, oxygen uptake expressed in $\mathrm{mL} \times \mathrm{kg}^{-1} \times \mathrm{min}^{-1}$; WR, work rate (watt); BM, body mass (kg); MET, metabolic equivalent of task.

Table 2.5 displays the corresponding WR value at the intersection of MET values (top row) and body mass (left column).

**Table 2.5** Metabolic equivalent of task (treadmill) to Watts (cycle ergometer) conversion depending on body mass according to the American College of Sports Medicine cycle equation [19] transformation

| BM (kg) | MET estimated according ACSM equation (treadmill ExT) | | | | | | | | | |
|---|---|---|---|---|---|---|---|---|---|---|
| | 5.0 | 6.0 | 7.0 | 8.0 | 9.0 | 10.0 | 11.0 | 12.0 | 13.0 | 14.0 |
| 60 | 57 | 76 | 95 | 114 | 133 | 153 | 172 | 191 | 210 | 229 |
| 64 | 61 | 81 | 102 | 122 | 142 | 163 | 183 | 203 | 224 | 244 |
| 68 | 65 | 86 | 108 | 130 | 151 | 173 | 194 | 216 | 238 | 259 |
| 72 | 69 | 92 | 114 | 137 | 160 | 183 | 206 | 229 | 252 | 275 |
| 76 | 72 | 97 | 121 | 145 | 169 | 193 | 217 | 241 | 266 | 290 |
| 80 | 76 | 102 | 127 | 153 | 178 | 203 | 229 | 254 | 280 | 305 |
| 84 | 80 | 107 | 133 | 160 | 187 | 214 | 240 | 267 | 294 | 320 |
| 88 | 84 | 112 | 140 | 168 | 196 | 224 | 252 | 280 | 308 | 336 |
| 92 | 88 | 117 | 146 | 175 | 205 | 234 | 263 | 292 | 322 | 351 |
| 96 | 92 | 122 | 153 | 183 | 214 | 244 | 275 | 305 | 336 | 366 |
| 100 | 95 | 127 | 159 | 191 | 222 | 254 | 286 | 318 | 349 | 381 |

The MET value obtained in the treadmill test should be reduced by 10–20% to estimate cycling performance in MET

BM, body mass; MET, metabolic equivalent of task; ACSM, American College of Sports Medicine

- MET to watt calculation by FRIEND-cycle ergometry equation [20] transformation

$$VO_2 = 1.74 \times \frac{WR \times 6.12}{BM} + 3.5$$

$$MET \times 3.5 = 10.65 \times WR/BM + 3.5$$

$$MET \times 3.5 - 3.5 = 10.65 \times \frac{WR}{BM}$$

$$BM \times (MET \times 3.5 - 3.5) = 10.65 \times WR$$

$$WR = BM \times (MET \times 3.5 - 3.5)/10.65 \tag{2.7}$$

where: $VO_2$, oxygen uptake ($mL \times kg^{-1} \times min^{-1}$); WR, work rate (watt); BM, body mass (kg); MET, metabolic equivalent of task.

The correlation between the MET value measured or estimated by the FRIEND-equation on the treadmill [21] and the load value on the bicycle depending on body weight is presented in Table 2.6.

- MET to watt calculation utilizing $VO_2$/WR slope method

**Table 2.6** MET's (treadmill) to Watts (cycle ergometer) conversion depending on body mass according to FRIEND-ergometry equation [20] transformation

| BM (kg) | MET measured or estimated according to FRIEND equation (treadmill) | | | | | | | | | |
|---|---|---|---|---|---|---|---|---|---|---|
| | 4.0 | 5.0 | 6.0 | 7.0 | 8.0 | 9.0 | 10.0 | 11.0 | 12.0 | 13.0 |
| 60 | 59 | 79 | 99 | 118 | 138 | 158 | 177 | 197 | 217 | 237 |
| 64 | 63 | 84 | 105 | 126 | 147 | 168 | 189 | 210 | 231 | 252 |
| 68 | 67 | 89 | 112 | 134 | 156 | 179 | 201 | 223 | 246 | 268 |
| 72 | 71 | 95 | 118 | 142 | 166 | 189 | 213 | 237 | 260 | 284 |
| 76 | 75 | 100 | 125 | 150 | 175 | 200 | 225 | 250 | 275 | 300 |
| 80 | 79 | 105 | 131 | 158 | 184 | 210 | 237 | 263 | 289 | 315 |
| 84 | 83 | 110 | 138 | 166 | 193 | 221 | 248 | 276 | 304 | 331 |
| 88 | 87 | 116 | 145 | 174 | 202 | 231 | 260 | 289 | 318 | 347 |
| 92 | 91 | 121 | 151 | 181 | 212 | 242 | 272 | 302 | 333 | 363 |
| 96 | 95 | 126 | 158 | 189 | 221 | 252 | 284 | 315 | 347 | 379 |
| 100 | 99 | 131 | 164 | 197 | 230 | 263 | 296 | 329 | 362 | 394 |

The MET value obtained in the treadmill test should be reduced by 10–20% to estimate cycling performance in MET

BM, body mass; MET, metabolic equivalent of task; FRIEND, Fitness Registry and the Importance of Exercise National Database

Another way to calculate ePWR is to use the relationship between the oxygen uptake increase and the external power (DVO2/DWR mL × min-1 × watt-1) in the cycle test. Therefore, it is enough to divide the corrected VO2 value assessed in the treadmill test by this relatively stable parameter in healthy people using the following formula [22]:

$$ePWR = \frac{(MET/1.1 - 1.8) \times BM \times 3.5}{9.6} \tag{2.8}$$

where: ePWR, estimated peak work rate with reduced MET by 10%; MET, estimated or measured metabolic equivalent of task; 1.1, 10% treadmill/cycle capacity correction factor according to Wasserman [23]; 1.8, energy expenditure in MET during unloaded cycling; BM, body mass (kg); 3.5, oxygen uptake ($mL \times kg^{-1} \times min^{-1}$) equivalent of 1 MET; 9.6, value of $\Delta VO_2/\Delta WR$ ($mL \times min^{-1} \times watt^{-1}$) [22].

The results of the conversion of MET to watt, depending on body weight, with a 10% reduction in functional capacity performance on the treadmill compared to the cycle ergometer, according to the $VO_2/WR$ slope method, are presented in Table 2.7

**Table 2.7** Metabolic equivalent of task to watt conversion depending on body mass—$VO_2$/WR slope method according to reference [22]

| BM (kg) | MET measured or estimated according to FRIEND-equation (treadmill) | | | | | | | | | |
|---|---|---|---|---|---|---|---|---|---|---|
| | 4.0 | 5.0 | 6.0 | 7.0 | 8.0 | 9.0 | 10.0 | 11.0 | 12.0 | 13.0 |
| 60 | 40 | 60 | 80 | 100 | 120 | 140 | 159 | 179 | 199 | 219 |
| 64 | 43 | 64 | 85 | 106 | 128 | 149 | 170 | 191 | 213 | 234 |
| 68 | 46 | 68 | 91 | 113 | 136 | 158 | 181 | 203 | 226 | 248 |
| 72 | 48 | 72 | 96 | 120 | 144 | 168 | 191 | 215 | 239 | 263 |
| 76 | 51 | 76 | 101 | 126 | 152 | 177 | 202 | 227 | 252 | 278 |
| 80 | 54 | 80 | 107 | 133 | 160 | 186 | 213 | 239 | 266 | 292 |
| 84 | 56 | 84 | 112 | 140 | 168 | 195 | 223 | 251 | 279 | 307 |
| 88 | 59 | 88 | 117 | 146 | 176 | 205 | 234 | 263 | 292 | 321 |
| 92 | 62 | 92 | 123 | 153 | 184 | 214 | 245 | 275 | 306 | 336 |
| 96 | 64 | 96 | 128 | 160 | 192 | 223 | 255 | 287 | 319 | 351 |
| 100 | 67 | 100 | 133 | 166 | 200 | 233 | 266 | 299 | 332 | 365 |

Watt is estimated after 10% reduction of MET
BM, body mass; MET, metabolic equivalent of task; FRIEND, Fitness Registry and the Importance of Exercise National Database

## 2.1.5 *Training $VO_2$ and MET Calculation Methods*

Two methods are applied to calculate training oxygen uptake ($TVO_2$) based on the measurement of peak oxygen consumption in the cardiopulmonary exercise test:

- %$VO_2R$—percentage of the oxygen uptake reserve method in which $TVO_2$ is calculated according to the formula:

$$TVO_2 = 0.01 \times Int \times \left(VO_{2peak} - VO_{2rest}\right) + VO_{2rest} \tag{2.9}$$

  where: $TVO_2$, training oxygen uptake ($mL \times kg^{-1} \times min^{-1}$); 0.01, decimal conversion factor for intensity: Int, desired intensity of training (%); $VO_{2peak}$, peak oxygen uptake ($mL \times kg^{-1} \times min^{-1}$); $VO_{2rest}$, resting oxygen uptake = 3.5 ($mL \times kg^{-1} \times min^{-1}$).
- %$VO_{2peak}$—of the peak oxygen uptake method:

$$TVO_2 = 0.01 \times Int \times VO_{2peak} \tag{2.10}$$

  where: $TVO_2$, training oxygen uptake ($mL \times kg^{-1} \times min^{-1}$); 0.01, decimal conversion factor for intensity: Int, desired training intensity (%); $VO_{2peak}$, peak oxygen uptake ($mL \times kg^{-1} \times min^{-1}$).

The calculated $TVO_2$ values according to the desired training intensity classes (Tables 2.1, 2.2 and 2.4) should be converted into the corresponding training control indicators: heart rate and/or work rate or may be the basis for programming the loads on the treadmill training.

The training MET (TMET) value, based on the estimated peak capacity expressed in MET and assessed in the treadmill test, is calculated using the following two methods:

- %MET-R—percent of the metabolic equivalent of task reserve method:

$$TMET = 0.01 \times Int \times \left(MET_{peak} - MET_{rest}\right) + MET_{rest}$$

$$TMET = 0.01 \times Int \times \left(MET_{peak} - 1\right) + 1 \quad (2.11)$$

where: TMET, training metabolic equivalent of task; 0.01, decimal conversion factor for intensity: Int, desired training intensity (%); $MET_{peak}$, peak metabolic equivalent of task; $MET_{rest}$, resting metabolic equivalent of task, $MET_{rest}=1$.

- $\%MET_{peak}$—percent of the peak metabolic equivalent of task method:

$$TMET = 0.01 \times Int \times MET_{peak} \quad (2.12)$$

where: TMET, training metabolic equivalent of task; 0.01, decimal conversion factor for intensity; Int, desired training intensity (%); $MET_{peak}$, peak metabolic equivalent of task.

### Treadmill Walking Programming

The following calculation based on absolute and relative VO2 has been developed:

$$1000 \times VO_2\left(L \times min^{-1}\right) = VO_2\left(mL \times min^{-1}\right)$$

$$VO_2 \frac{\left(mL \times min^{-1}\right)}{BM\,(kg)} = VO_2\left(mL \times kg^{-1} \times min^{-1}\right)$$

$$VO_2 \frac{\left(mL \times kg^{-1} \times min^{-1}\right)}{3.5\left(mL \times kg^{-1} \times min^{-1}\right)} = MET \quad (2.13)$$

where: $VO_2$, oxygen uptake in absolute and relative units; BM, body mass; 3.5 $VO_2$ ($mL \times kg^{-1} \times min^{-1}$) equivalent of 1 MET; MET, metabolic equivalent of task.

Calculation of treadmill training parameters (speed and grade) in "manual mode" is performed in the following three steps:

- determining the "optimal treadmill training speed" using a ramp test, based on the patient's report.
- calculation of TMET for desired intensity of training,
- calculation or respective treadmill inclination (Grade) from chart Speed-Grade-MET (Tables 2.8 or 2.9 depending on the method of estimation the MET on the treadmill)

Calculation of treadmill Grade based on Speed and MET by ACSM equation for treadmill [19] (Eq. 1.3) according to the following steps:

$$VO_2 = (S \times 1.66) + (0.3 \times S \times G) + 3.5$$

$$MET \times 3.5 = (S \times 1.66) + (0.3 \times S \times G) + 3.5$$

$$(MET \times 3.5) - (S \times 1.66) - 3.5 = (0.3 \times S \times G)$$

$$G = [(MET \times 3.5) - (S \times 1.66) - 3.5]/(0.3 \times S) \tag{2.14}$$

where: $VO_2$, oxygen uptake ($mL \times kg^{-1} \times min^{-1}$); S, speed (km/h); G, grade (%); MET, metabolic equivalent of task.

**Table 2.8** Speed-Grade-MET correlation on treadmill according to the American College of Sports Medicine equation [19]

| Grade (%) | Speed (km/h) | | | | | | | | | |
|---|---|---|---|---|---|---|---|---|---|---|
| | 2.5 | 3.0 | 3.5 | 4.0 | 4.5 | 5.0 | 5.5 | 6.0 | 6.5 | 7.0 |
| 0 | 2.2 | 2.4 | 2.7 | 2.9 | 3.1 | 3.4 | 3.6 | 3.9 | 4.1 | 4.3 |
| 2 | 2.6 | 2.9 | 3.3 | 3.6 | 3.9 | 4.2 | 4.6 | 4.9 | 5.2 | 5.5 |
| 4 | 3.0 | 3.5 | 3.9 | 4.3 | 4.7 | 5.1 | 5.5 | 5.9 | 6.3 | 6.7 |
| 6 | 3.5 | 4.0 | 4.5 | 5.0 | 5.5 | 6.0 | 6.4 | 6.9 | 7.4 | 7.9 |
| 8 | 3.9 | 4.5 | 5.1 | 5.6 | 6.2 | 6.8 | 7.4 | 8.0 | 8.6 | 9.1 |
| 10 | 4.3 | 5.0 | 5.7 | 6.3 | 7.0 | 7.7 | 8.3 | 9.0 | 9.7 | 10.3 |
| 12 | 4.8 | 5.5 | 6.3 | 7.0 | 7.8 | 8.5 | 9.3 | 10.0 | 10.8 | 11.5 |
| 14 | 5.2 | 6.0 | 6.9 | 7.7 | 8.5 | 9.4 | 10.2 | 11.1 | 11.9 | 12.7 |
| 16 | 5.6 | 6.5 | 7.5 | 8.4 | 9.3 | 10.2 | 11.2 | 12.1 | 13.0 | 13.9 |
| 18 | 6.0 | 7.1 | 8.1 | 9.1 | 10.1 | 11.1 | 12.1 | 13.1 | 14.1 | 15.1 |
| 20 | 6.5 | 7.6 | 8.7 | 9.8 | 10.9 | 12.0 | 13.0 | 14.1 | 15.2 | 16.3 |
| 22 | 6.9 | 8.1 | 9.3 | 10.4 | 11.6 | 12.8 | 14.0 | 15.2 | 16.4 | 17.5 |
| 24 | 7.3 | 8.6 | 9.9 | 11.1 | 12.4 | 13.7 | 14.9 | 16.2 | 17.5 | 18.7 |

To determine the inclination of the treadmill, find the calculated training MET value in the column indicated by the optimal training speed of the treadmill (top row) and use the inclination in the left column at the level of the row indicated by MET

**Table 2.9** Speed-Grade-MET correlation on treadmill according to FRIEND-equation [21]

| Grade (%) | Speed (km/h) | | | | | | | | | |
|---|---|---|---|---|---|---|---|---|---|---|
| | 2.5 | 3.0 | 3.5 | 4.0 | 4.5 | 5.0 | 5.5 | 6.0 | 6.5 | 7.0 |
| 0 | 3.0 | 3.4 | 3.8 | 4.2 | 4.6 | 5.0 | 5.5 | 5.9 | 6.3 | 6.7 |
| 2 | 3.2 | 3.7 | 4.1 | 4.5 | 5.0 | 5.4 | 5.9 | 6.3 | 6.7 | 7.2 |
| 4 | 3.4 | 3.9 | 4.4 | 4.8 | 5.3 | 5.8 | 6.3 | 6.8 | 7.2 | 7.7 |
| 6 | 3.6 | 4.1 | 4.6 | 5.1 | 5.7 | 6.2 | 6.7 | 7.2 | 7.7 | 8.2 |
| 8 | 3.8 | 4.3 | 4.9 | 5.4 | 6.0 | 6.6 | 7.1 | 7.7 | 8.2 | 8.8 |
| 10 | 4.0 | 4.6 | 5.1 | 5.7 | 6.3 | 6.9 | 7.5 | 8.1 | 8.7 | 9.3 |
| 12 | 4.2 | 4.8 | 5.4 | 6.0 | 6.7 | 7.3 | 7.9 | 8.6 | 9.2 | 9.8 |
| 14 | 4.3 | 5.0 | 5.7 | 6.3 | 7.0 | 7.7 | 8.3 | 9.0 | 9.7 | 10.3 |
| 16 | 4.5 | 5.2 | 5.9 | 6.6 | 7.3 | 8.1 | 8.8 | 9.5 | 10.2 | 10.9 |
| 18 | 4.7 | 5.5 | 6.2 | 6.9 | 7.7 | 8.4 | 9.2 | 9.9 | 10.7 | 11.4 |
| 20 | 4.9 | 5.7 | 6.5 | 7.2 | 8.0 | 8.8 | 9.6 | 10.4 | 11.1 | 11.9 |
| 22 | 5.1 | 5.9 | 6.7 | 7.5 | 8.4 | 9.2 | 10.0 | 10.8 | 11.6 | 12.5 |
| 24 | 5.3 | 6.1 | 7.0 | 7.8 | 8.7 | 9.6 | 10.4 | 11.3 | 12.1 | 13.0 |

To determine the inclination of the treadmill, find the calculated training MET value in the column indicated by the optimal training speed of the treadmill (top row) and use the inclination in the left column at the level of the row indicated by MET

Table 2.8 illustrates the relationship between the mechanical parameters of the treadmill (speed and grade) and MET calculated according to the ACSM.

Transformation of the FRIEND-equation [21], used for estimation of $MET_{peak}$ or measured, for calculated treadmill slope after determination of the optimal speed and training MET:

$$VO_2 = 16.66 \times \mathrm{S} \times (0.17 + G \times 0.0079) + 3.5$$

$$\mathrm{MET} \times 3.5 = 16.66 \times \mathrm{S} \times (0.17 + G \times 0.0079) + 3.5$$

$$(MET \times 3.5 - 3.5) = (16.66 \times S) \times (0.17 + G \times 0.0079)$$

$$(MET \times 3.5 - 3.5)/(16.66 \times S) = (0.17 + G \times 0.0079)$$

$$((MET \times 3.5 - 3.5))/((16.66 \times S)) - 0.17 = (G \times 0.0079)$$

$$G = [(MET \times 3.5 - 3.5)/(16.66 \times S) - 0.17]/0.0079 \qquad (2.15)$$

where: $VO_2$, oxygen uptake ($mL \times kg^{-1} \times min^{-1}$); S, speed (km/h); G, grade (%); MET, metabolic equivalent of task.

The dependence of MET on the speed and inclination of the treadmill according to the FRIEND -equation is presented in Table 2.9.

## 2.1.6 Subjective Methods in Intensity Programming

In training programming and monitoring, the following two subjective methods are commonly used as a primary method or combined with objective indicators of exercise intensity (HR, WR, $VO_2$, and MET) [9]:

- rate of perceived exertion in Borg scales
- Talk Test.

**Rate of Perceived Exertion—Borg Scales**

The subjective assessment of the rate of perceived exertion (RPE) in cardiac rehabilitation is used both in qualifying an exercise test as maximal, i.e., $^{3}$18 in the RPE scale or $^{3}$8 in the CR10 scale [11] or submaximal, as well as in programming and monitoring exercise training, especially when the HR monitoring method is of limited application [24, 25].

Commonly used RPE assessment tools are the Borg scales [26]:

- The Rate of Perceived Exertion Scale (Borg RPE Scale) (Fig. 2.11)
- The Borg Category Ratio Scale (Borg CR10 Scale) (Fig. 2.12).

In the Borg Rating of Perceived Exertion (RPE) scale, individuals assess their level of perceived effort using a 15-point scale that ranges from 6 to 20. This scale effectively captures the overall psychophysical impact of the exercise being evaluated. Additionally, the Borg CR10 scale, introduced in 1982, utilizes a simpler 10-point scale from 0 to 10 to classify perceived effort. Research has shown a high degree of equivalence and interchangeability between the two scales. The correlation between the Borg RPE and Borg CR10 scales during both incremental and interval exercise has been established through the equations outlined below [27]. Incremental:

$$BorgCR10 = 0.02 \times (BorgRPE)^2 + 0.1659 \times (BorgRPE) - 1.3221$$

$$BorgRPE = -0.0615 \times (BorgCR10)^2 + 1.9065 \times (BorgCR10) + 6.0806$$

Interval:

$$BorgCR10 = 0.0335 \times (BorgRPE)^2 - 0.142 \times (BorgRPE) + 0.3372$$

$$BorgRPE = -0.047 \times (BorgCR10)^2 + 1.7007 \times (BorgCR10) + 6.4579$$

**Fig. 2.11** The Borg RPE Scale® © Rating of Perceived Exertion. Gunnar Borg. 1970, 1998, 2017. Scale printed with permission. The scale and full instruction can be obtained through Borg Perception www.borgperception.se

**Borg RPE Scale**

| | |
|---|---|
| 6 | No exertion at all |
| 7 | |
| | Extremely light |
| 8 | |
| 9 | Very light |
| 10 | |
| 11 | Light |
| 12 | Moderate |
| 13 | Somewhat hard |
| 14 | |
| 15 | Hard |
| 16 | |
| 17 | Very hard |
| 18 | |
| 19 | Extremely hard |
| 20 | Maximal exertion |

RPE correlates with relative $VO_2$, HR and work rate values [28, 29]. Using the individual linear relationship between RPE and $VO_2$ assessed in a submaximal exercise test, the predicted $VO_{2max}$ or $VO_{2peak}$ is determined. The RPE: $VO_2$ data pairs needed for calculation of the linear regression equation and for extrapolation of $VO_{2max}$ for RPE = 20 or $VO_{2peak}$ for RPE = 19 is acquired in two modes [28]:

- estimation mode—RPE: $VO_2$ recorded at every stage of submaximal (i.e. to RPE = 15 or RPE = 17) graded exercise test [30]
- production mode—RPE: $VO_2$ are recorded from a perceptually regulated exercise test (PRET) in which the subject regulates the load in active 2–4 min steps according to the protocol of corresponding loads (e.g. RPE 9, 11, 13, 15, 17) [31].

In the CARE CR study the relation between the RPE and the load at the ventilatory threshold has been documented [32].

**Borg CR10 Scale**

| | | |
|---|---|---|
| **0** | **Nothing at all** | |
| **0.3** | | |
| **0.5** | **Extremely weak** | **Just noticeable** |
| **0.7** | | |
| **1** | **Very weak** | |
| **1.5** | | |
| *2* | **Weak** | **Light** |
| **2.5** | | |
| **3** | **Moderate** | |
| **4** | | |
| **5** | **Strong** | **Heavy** |
| **6** | | |
| **7** | **Very strong** | |
| **8** | | |
| **9** | | |
| **10** | **Extremely strong** | **"Maximal"** |
| **11** | | |
| ⋮ | | |
| • | **Absolute maximum** | **Highest possible** |

**Fig. 2.12** The Borg CR10 Scale® (CR10) ©. Gunnar Borg, 1982, 1998, 2004. Scale printed with permission. The scale can be obtained through Borg Perception www.borgperception.se

Two months of exercise training based on RPE resulted in an increase in physical performance and improvement in weight reduction, lipids profile, and blood pressure [33].

Likewise, positive effects in patients undergoing cardiac surgery have been described [24].

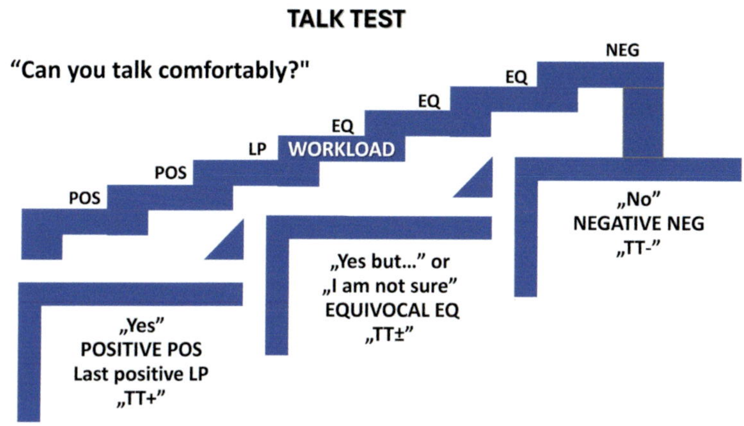


**Fig. 2.13** Diagram of the Talk Test Procedure with a description of the abbreviations used, by the authors

## Talk Test

The Talk Test is commonly used in cardiac rehabilitation labs based on the findings that efforts at the aerobic threshold level affects comfortable talk, and exercising at intensity beyond the respiratory compensation does not allow to speak.

The existing talk test protocols overview can be found else where [34–39]. The schematic TT procedure is illustrated in Fig. 2.13.

During the exercise test at the end of each stage or at specific intervals (in case of the ramp test), the subject is asked to read aloud a text of 30–90 words and is asked whether he/she can talk comfortably. Three types of responses corresponding to workloads determine the TT stages:

- "Yes"—positive stage (TT+) with indexes of last positive (LP) load (VO2, HR, WR or treadmill parameters: speed and grade for VO2 estimation)
- "Yes but…" or "I am not sure"—equivocal stage (TT±)
- "No" negative stage (TT-), which is an indication to end the test

Studies have shown the moderate to high correlation of VO2, HR, and WR at the VT1 and VT2 threshold determined in CPET and these indices in TT stages in healthy sedentary, active, elderly, and rehabilitated patients with CVD [37, 40–43]. A positive stage (TT+) usually correlates with loads below VT1, and the first degree rated as equivocal (TT±) indicates a threshold load. A negative response (TT-) suggests that the respiratory compensation threshold has been reached [40].

The systematic review of nine qualified studies covering 425 mainly CVD patients concluded that TT is suitable as an alternative method to assess performance and to guide the training intensity in patients with cardiovascular diseases. Aerobic training is recommended at an intensity corresponding to the "TT+" stage when the patient can still speak comfortably [43]. Guidelines recommend using the Talk Test and Borg should only as an additional tool to an objective exercise intensity determination.

**AuthorContribution Competing Interests** The authors have no conflicts of interest to declare that are relevant to the content of this chapter.

## References

1. Ambrosetti M, Abreu A, Corrà U, Davos CH, Hansen D, Frederix I, et al. Secondary prevention through comprehensive cardiovascular rehabilitation: from knowledge to implementation. 2020 update. A position paper from the secondary prevention and rehabilitation section of the European association of preventive cardiology. Eur J Prev Cardiol. 2021;28:460–495. https://doi.org/10.1177/2047487320913379.
2. Lang JJ, Prince SA, Merucci K, Cadenas-Sanchez C, Chaput JP, Fraser BJ, et al. Cardiorespiratory fitness is a strong and consistent predictor of morbidity and mortality among adults: an overview of meta-analyses representing over 20.9 million observations from 199 unique cohort studies. Br J Sports Med. 2024;58:556–566. https://doi.org/10.1136/bjsports-2023-107849. PMID: 38599681; PMCID: PMC11103301.
3. Thompson PD, Stern MP, Williams P, Duncan K, Haskell WL, Wood PD. Death during jogging or running. A study of 18 cases. JAMA. 197921;242:1265–7. PMID: 480538.
4. Albert CM, Mittleman MA, Chae CU, Lee IM, Hennekens CH, Manson JE. Triggering of sudden death from cardiac causes by vigorous exertion. N Engl J Med. 2000;343:1355–61. https://doi.org/10.1056/NEJM200011093431902.
5. Franklin BA. Cardiovascular events associated with exercise. The risk-protection paradox J Cardiopulm Rehabil. 2005;25:189–95. https://doi.org/10.1097/00008483-200507000-00001.
6. AACVPR Stratification Algorithm for Risk of Event https://registry.dev.aacvpr.org/Documents/AACVPR%20Risk%20Stratification%20Algorithm_June2012.pdf
7. Silva AK, Barbosa MP, Bernardo AF, Vanderlei FM, Pacagnelli FL, Vanderlei LC. Cardiac risk stratification in cardiac rehabilitation programs: a review of protocols. Rev Bras Cir Cardiovasc. 2014;29:255–65. https://doi.org/10.5935/1678-9741.20140067.
8. Garber CE, Blissmer B, Deschenes MR, Franklin BA, Lamonte MJ, Lee IM, et al. American college of sports medicine. American college of sports medicine position stand. Quantity and quality of exercise for developing and maintaining cardiorespiratory, musculoskeletal, and neuromotor fitness in apparently healthy adults: guidance for prescribing exercise. Med Sci Sports Exerc. 2011;43:1334–59. https://doi.org/10.1249/MSS.0b013e318213fefb.
9. Hansen D, Abreu A, Ambrosetti M, Cornelissen V, Gevaert A, Kemps H, et al. Exercise intensity assessment and prescription in cardiovascular rehabilitation and beyond: why and how: a position statement from the Secondary Prevention and Rehabilitation Section of the European Association of Preventive Cardiology. Eur J Prev Cardiol. 2022;29:230–45. https://doi.org/10.1093/eurjpc/zwab007.
10. Binder RK, Wonisch M, Corra U, Cohen-Solal A, Vanhees L, et al. Methodological approach to the first and second lactate threshold in incremental cardiopulmonary exercise testing. Eur J Cardiovasc Prev Rehabil. 2008;15:726–34. https://doi.org/10.1097/HJR.0b013e328304fed4.
11. Mezzani A, Hamm LF, Jones AM, McBride PE, Moholdt T, Stone JA, et al. European association for cardiovascular prevention and rehabilitation; American association of cardiovascular and pulmonary rehabilitation; Canadian association of cardiac rehabilitation. Aerobic exercise intensity assessment and prescription in cardiac rehabilitation: a joint position statement of the European association for cardiovascular prevention and rehabilitation, the American association of cardiovascular and pulmonary rehabilitation and the Canadian association of cardiac rehabilitation. Eur J Prev Cardiol. 2013;20:442–467. https://doi.org/10.1177/2047487312460484.

12. Pelliccia A, Sharma S, Gati S, Bäck M, Börjesson M, Caselli S, et al. ESC Scientific document group. 2020 ESC guidelines on sports cardiology and exercise in patients with cardiovascular disease. Eur Heart J. 2021;42:17–96. https://doi.org/10.1093/eurheartj/ehaa605.
13. Franklin BA, Arena R, Kaminsky LA, Peterman JE, Kokkinos P, Myers J. Maximizing the cardioprotective benefits of exercise with age-, sex-, and fitness-adjusted target intensities for training. Eur J Prev Cardiol. 2022;29:e1–3. https://doi.org/10.1093/eurjpc/zwaa094.
14. Karvonen MJ, Kentala E, Mustala O. The effects of training on heart rate; a longitudinal study. Ann Med Exp Biol Fenn. 1957;35:307–15 PMID: 13470504.
15. Swain DP, Leutholtz BC. Heart rate reserve is equivalent to %VO2 reserve, not to %VO2max. Med Sci Sports Exerc. 1997;29:410–4. https://doi.org/10.1097/00005768-199703000-00018.
16. Brawner CA, Keteyian SJ, Ehrman JK. The relationship of heart rate reserve to VO2 reserve in patients with heart disease. Med Sci Sports Exerc. 2002;34:418–22. https://doi.org/10.1097/00005768-200203000-00006.
17. Chaudhry S, Arena R, Bhatt DL, Verma S, Kumar N. A practical clinical approach to utilize cardiopulmonary exercise testing in the evaluation and management of coronary artery disease: a primer for cardiologists. Curr Opin Cardiol. 2018;33:168–77. https://doi.org/10.1097/HCO.0000000000000494.
18. Williams RP, Manou-Stathopoulou V, Redwood SR, Marber MS. Warm-up Angina': harnessing the benefits of exercise and myocardial ischaemia. Heart. 2014;100:106–14. https://doi.org/10.1136/heartjnl-2013-304187.
19. Glass S, Dwyer G. ACSM's metabolic calculations handbook. Lippincott Williams & Wilkins 2007; p 22
20. Kokkinos P, Kaminsky LA, Arena R, Zhang J, Myers J. A new generalized cycle ergometry equation for predicting maximal oxygen uptake: the fitness registry and the importance of exercise national database (FRIEND). Eur J Prev Cardiol. 2018;25:1077–82. https://doi.org/10.1177/2047487318772667.
21. Kokkinos P, Kaminsky LA, Arena R, Zhang J, Myers J. New generalized equation for predicting maximal oxygen uptake (from the fitness registry and the importance of exercise national database). Am J Cardiol. 2017;120:688–92. https://doi.org/10.1016/j.amjcard.2017.05.037.
22. Saengsuwan J, Nef T, Hunt KJ. A method for predicting peak work rate for cycle ergometer and treadmill ramp tests. Clin Physiol Funct Imaging. 2017;37:610–4. https://doi.org/10.1111/cpf.12344.
23. Wasserman K. Diagnosing cardiovascular and lung pathophysiology from exercise gas exchange. Chest. 1997;112:1091–101. https://doi.org/10.1378/chest.112.4.1091.
24. Zanettini R, Centeleghe P, Ratti F, Benna S, Di Tullio L, Sorlini N. Training prescription in patients on beta-blockers: percentage peak exercise methods or self-regulation? Eur J Prev Cardiol. 2012;19:205–12. https://doi.org/10.1177/1741826711398823.
25. Tabet JY, Meurin P, Teboul F, Tartiere JM, Weber H, Renaud N, et al. Determination of exercise training level in coronary artery disease patients on beta blockers. Eur J Cardiovasc Prev Rehabil. 2008;15:67–72. https://doi.org/10.1097/HJR.0b013e3282eff61f.
26. Borg GA. Psychophysical bases of perceived exertion. Med Sci Sports Exerc. 1982;14:377–81 PMID: 7154893.
27. Arney BE, Glover R, Fusco A, Cortis C, de Koning JJ, van Erp T, et al. Comparison of rating of perceived exertion scales during incremental and interval exercise. Kinesiology. 2019;51:150–157. Retrieved from https://hrcak.srce.hr/ojs/index.php/kinesiology/article/view/6882
28. Coquart JB, Garcin M, Parfitt G, Tourny-Chollet C, Eston RG. Prediction of maximal or peak oxygen uptake from ratings of perceived exertion. Sports Med. 2014;44:563–78. https://doi.org/10.1007/s40279-013-0139-5. PMID: 24519666.
29. Eston RG, Thompson M. Use of ratings of perceived exertion for predicting maximal work rate and prescribing exercise intensity in patients taking atenolol. Br J Sports Med. 1997;31:114–9. https://doi.org/10.1136/bjsm.31.2.114.
30. Coquart JB, Lemaire C, Dubart AE, Douillard C, Luttenbacher DP, Wibaux F, Garcin M. Prediction of peak oxygen uptake from sub-maximal ratings of perceived exertion elicited during a graded exercise test in obese women. Psychophysiology. 2009;46:1150–3. https://doi.org/10.1111/j.1469-8986.2009.00863.x.

31. Eston R, Lambrick D, Sheppard K, Parfitt G. Prediction of maximal oxygen uptake in sedentary males from a perceptually regulated, sub-maximal graded exercise test. J Sports Sci. 2008;26:131–9. https://doi.org/10.1080/02640410701371364. PMID: 17852665.
32. Nichols S, Engin B, Carroll S, Buckley J, Ingle L. Ratings of perceived exertion at the ventilatory anaerobic threshold in people with coronary heart disease: a CARE CR study. Ann Phys Rehabil Med. 2021;64: 101462. https://doi.org/10.1016/j.rehab.2020.101462.292.
33. Parfitt G, Evans H, Eston R. Perceptually regulated training at RPE13 is pleasant and improves physical health. Med Sci Sports Exerc. 2012;44:1613–8. https://doi.org/10.1249/MSS.0b013e31824d266e.
34. Foster C, Porcari JP, Doro K, Dubiel J, Engen M, Kolman D, et al. Exercise prescription when there is no exercise test: the Talk Test. Kinesiology. 2018;50:333–48.
35. Reed JL, Pipe AL. The talk test: a useful tool for prescribing and monitoring exercise intensity. Curr Opin Cardiol. 2014;29:475–80. https://doi.org/10.1097/HCO.0000000000000097.
36. Nielsen SG, Buus L, Hage T, Olsen H, Walsøe M, Vinther A. The graded cycling test combined with the talk test is reliable for patients with ischemic heart disease. J Cardiopulm Rehabil Prev. 2014;34:276–80. https://doi.org/10.1097/HCR.0000000000000067.
37. Sørensen L, Larsen KSR, Petersen AK. Validity of the talk test as a method to estimate ventilatory threshold and guide exercise intensity in cardiac patients. J Cardiopulm Rehabil Prev. 2020;40:330–4. https://doi.org/10.1097/HCR.0000000000000506.
38. Quinn TJ, Coons BA. The Talk Test and its relationship with the ventilatory and lactate thresholds. J Sports Sci. 2011;29:1175–82. https://doi.org/10.1080/02640414.2011.585165.
39. Lyon E, Menke M, Foster C, Porcari JP, Gibson M, Bubbers T. Translation of incremental talk test responses to steady-state exercise training intensity. J Cardiopulm Rehabil Prev. 2014;34:271–5. https://doi.org/10.1097/HCR.0000000000000069.
40. Ballweg J, Foster C, Porcari J, Haible S, Aminaka N, Mikat RP. Reliability of the talk test as a surrogate of ventilatory and respiratory compensation thresholds. J Sports Sci Med. 2013;12:610–1. PMID: 24137081; PMCID: PMC3772610.
41. Rodríguez-Marroyo JA, Villa JG, García-López J, Foster C. Relationship between the talk test and ventilatory thresholds in well-trained cyclists. J Strength Cond Res. 2013;27:1942–9. https://doi.org/10.1519/JSC.0b013e3182736af3. PMID: 23007491.
42. Althoff A, Vieira AM, Silveira LSD, Benetti M, Karsten M. Aerobic exercise prescription in cardiac rehabilitation based on heart rate from talk test stages and 6-minute walk test. Arq Bras Cardiol. 2023;120: e20230086. https://doi.org/10.36660/abc.20230086.PMID:37820173; PMCID:PMC10519355.
43. Vieira AM, Martins EM, Althoff A, Rech DA, Ribeiro GDS, Matte DL, Karsten M. Application and measurement properties of the talk test in cardiopulmonary patients: a systematic review. Rev Cardiovasc Med. 2022;23:225. https://doi.org/10.31083/j.rcm2307225.PMID:39076925; PMCID:PMC11266803.

# Chapter 3
# Exercise Training Protocols

**Adam Staron, Jerzy Rybicki, and Adré Rossouw**

**Abstract** Exercise training remains the mainstay of the contemporary comprehensive cardiac rehabilitation program, offering numerous beneficial effects for cardiovascular health. The most common aerobic exercise training protocols endorsed in cardiac rehabilitation facilities include moderate-intensity continuous training complemented by resistance, flexibility, neuromotor exercises, and functional training. However, with the numerous studies exploring the exercise-induced preconditioning phenomenon, there is revitalization of high-intensity interval training as well, and attempts are being made to modify existing interval training protocols, e.g., pyramidal training or interval-to-continuous training. The optimal approach to exercise training programming should be tailored to efficacy and safety and tailored to patients' needs. The delivery of exercise stimulus during a supervised exercise training program should be sufficient to evoke cardio protection and to allow for gradual training progression.

**Keywords** Endurance training · Strength training · Flexibility training · Balance training · Continuous training · Interval training · Interval-to-continuous training · Pyramidal training

A. Staron (✉)
The Upper Silesian Medical Center, Katowice, Poland
e-mail: astaron1973@wp.pl

J. Rybicki
"REPTY" Upper Silesian Rehabilitation Center Tarnowskie Góry Poland, Silesian Center for Heart Disease, Zabrze, Poland
e-mail: j_rybicki@cyberia.pl

A. Rossouw
Clinical Exercise Physiologist Cardia Clinic, Jabriya, Kuwait

 
J. Rybicki and A. Staron (eds.), *Exercise Programming for Cardiac Patients*,
https://doi.org/10.1007/978-3-031-88475-7_3

## 3.1 Continuous Training Models

Continuous training is the commonly recommended exercise training type according to international guidelines, and it typically ranges from low to moderate intensity. Protocols usually start with a 3–5 min warm-up at an intensity that is below the first ventilatory threshold, followed by the main phase of 20–40 (50) min, and finish with a 3–5 min cool-down phase [1]. Training intensity depends on the clinical condition and the risk of exercise-induced cardiac events. In continuous training on a cycle ergometer or on a treadmill, intensity is programmed based on the peak tolerable work (peak work rate PWR) in watts for the bicycle and MET for the treadmill, with stable and repetitive load applied. Heart rate-controlled continuous training is achievable due to the development of heart rate-load feedback, enforcing the attainment of training heart rate (THR) after the warm-up phase and maintaining the desired heart rate (HR) range within the main phase of the training. Feasibility of the heart rate-controlled mode refers to continuous training only. Detailed characteristics of continuous training on leg cycle ergometers and treadmills have been described below.

### *3.1.1 Continuous Watt-Controlled Training on Cycle Ergometer*

- dedicated to patients with low and good physical capacity,
- should be preceded by an exercise test on a cycle ergometer with aim of determining peak work rate,
- training parameters include a frequency 5 per week for the residential program and 2–3 per week for outpatient cardiac rehabilitation,
- training intensity ranges from low (low-intensity continuous training LICT) to moderate-vigorous (moderate-intensity continuous training MICT), depending on functional assessment level and risk stratification (30–70% PWR, 30–70% heart rate reserve (HRR), rate of perceived exertion (RPE): 9–11 to 12–14),
- training duration 30–60 min, depending on the intensity applied
- training is progressed by increased time, preceding an increase in exercise intensity.

### *3.1.2 Continuous MET-Controlled Training on Treadmill*

- dedicated to patients with low and good functional capacity
- should be preceded by exercise testing on a treadmill, optimally utilizing ramp protocol

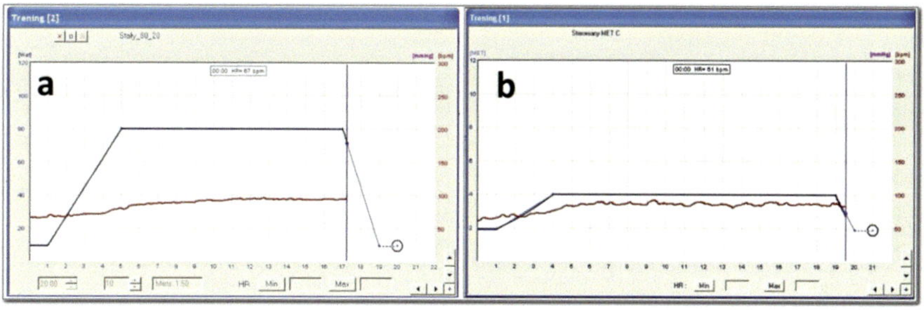


**Fig. 3.1** **a** load-controlled continuous training on a cycle ergometer and **b** on the treadmill (with permission from reference 2)

- training parameters include a frequency of 5 per week for residential programs and 2–3 per week for outpatient,
- training intensity ranges from low to vigorous (30–70% METpeak, 30–70% HRR, RPE 9–11 to 12–14),
- training duration 30–60 min, depending on training intensity
- training progression typically includes prolonging exercise time, followed by up-titration of exercise intensity.
- programming of Watt or MET-controlled continuous training is possible nowadays due to the advancement of cardiac rehabilitation software systems, which requires peak Watt or peak MET values, targeted intensity (%), and training time for optimal load or speed calculation. The example of the load-controlled continuous approach is presented in Fig. 3.1.

### *3.1.3 Heart Rate-Controlled Continuous Training on Cycle Ergometer or Treadmill*

- is preferred for patients with good functional capacity and normal chronotropic response
- should be preceded by a ramp exercise test on the same modality as is planned for training
- parameters include frequency 5 per week for residential program, 2–3 per week for outpatient,
- training intensity applied ranges from moderate to vigorous depending on functional capacity and risk stratification (40–70% HRR, RPE to 12–14),
- training duration of 30–60 min
- training progression depends on RPE declared, load response for imposed heart rate, clinical condition and typically is performed by prolongation of exercise time with further increase of exercise intensity.
- The example of heart rate-controlled continuous training is presented in Fig. 3.2.

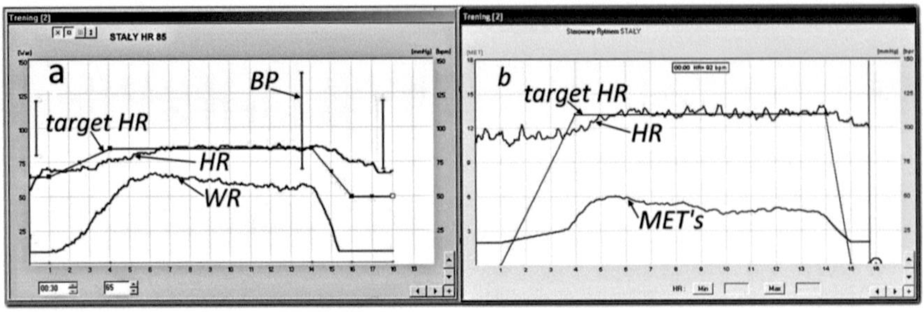


**Fig. 3.2** **a** heart rate-controlled continuous training on a leg ergometer, **b** treadmill heart rate-controlled continuous training (with permission from reference 2) BP, blood pressure; HR, heart rate: WR, work rate; METs, metabolic equivalent of tasks

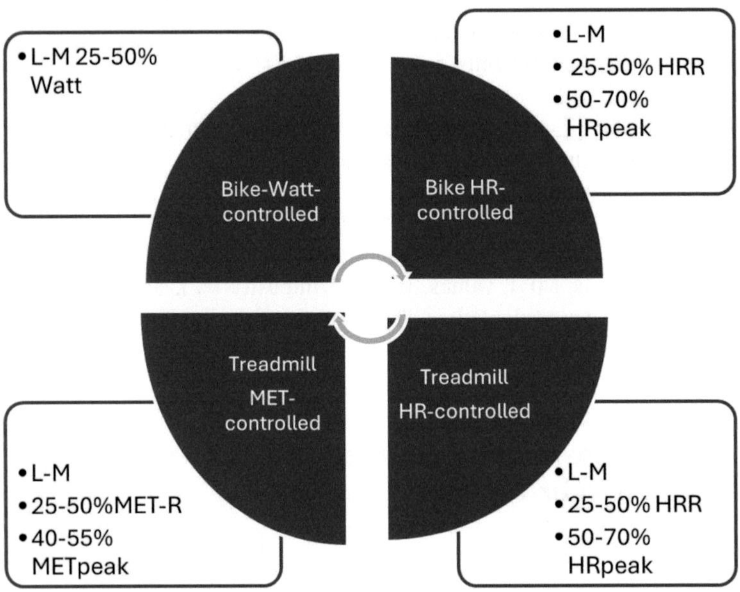


**Fig. 3.3** Continuous training at low-to-moderate intensity, *Source* by the authors HR, heart rate; HRR, heart rate reserve; L-M, low-to-moderate; MET, metabolic equivalent of task; METR, reserve of metabolic equivalents of task

Continuous exercise training can be the primary type of aerobic exercise protocol or can be performed as an initial stage in transitional exercise programs, e.g., in exercise training models suggested by the Mayo Clinic [1] and the European Association of Preventive Cardiology [2]. The principle of continuous exercise training has been demonstrated in Figs. 3.3 and 3.4 Continuous exercise training can be the main type of aerobic exercise protocol or can be performed as an initial stage in transitional exercise programs, e.g., in exercise training models suggested by Mayo

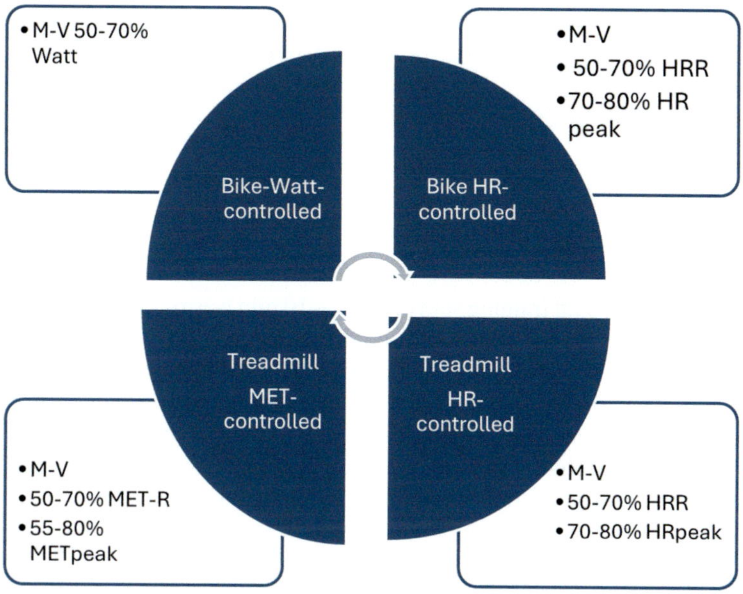


**Fig. 3.4** Continuous training at moderate-to-vigorous intensity, *Source* by the authors HR, heart rate; HRR, heart rate reserve; MET, metabolic equivalent of task; MET-R, reserve of metabolic equivalents of task; M-V, moderate-to-vigorous

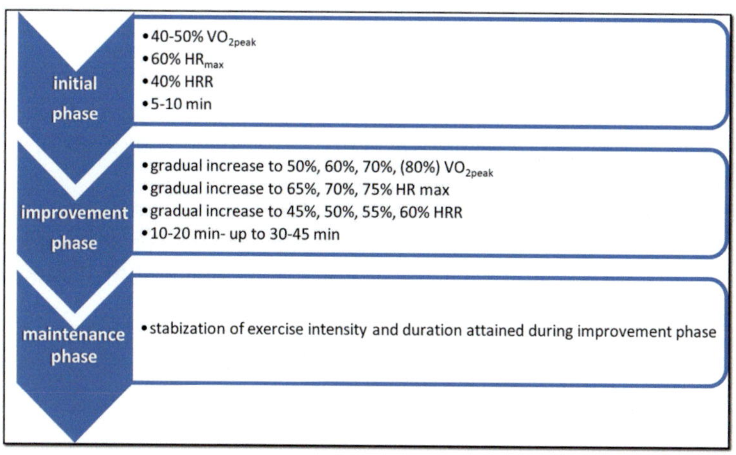


**Fig. 3.5** Progressive model of continuous training [3] HRR, heart rate reserve; HRmax, maximal heart rate; VO2 peak, peak oxygen uptake

Clinic [1] and the European Association of Preventive Cardiology [2]. The principle of continuous exercise training has been demonstrated in Figs. 3.3 and 3.4.

Vanhees and colleagues suggested phasic model of endurance training progression as presented in Fig. 3.5 [3].

## 3.2 Interval Training Models

Load-controlled interval training starts with a 3–10 min warm-up phase, followed by active load cycles typically at moderate, moderate-to-vigorous, or vigorous intensity and duration from 10 s to 4 min, separated by recovery periods. The training ends with a 3–10 min cool-down phase. Interval training is recommended for patients with heart failure, peripheral arterial disease, intermittent claudication, and good physical capacity [3]. Interval mode is typically performed on cycle leg ergometers; however, training on a treadmill is feasible by utilizing longer active phases. Training parameters include:

- frequency 5 per week for residential programs, 2–3 per week for outpatient
- training intensity on the bike ranges from moderate to vigorous (55–90% PWR, 50–80% MSEC for training according to Meyer, 50–80% HRR, RPE 12 to 15) for active phases and for 10–15 Watts for passive recovery phases, 30–50% PWR for active recovery phases
- training intensity on the treadmill ranges from moderate to vigorous for active phases (55–90% MET-R, 50–90% HRR), 2 MET (V–2 km/h, incl—0) for passive recovery, 30–55% MET-R for active recovery
- training duration 30-60 min including 3–10 min of warm-up, active phases of 10–240 s (bike), 1–4 min (treadmill), recovery phases of 15–180 s (bike), 1–3 min (treadmill)
- training progression depends on the declared RPE value, chronotropic response for a given load, clinical picture, and risk level and is typically implemented through the increase in the number of load cycles, followed by the increase in training intensity.

### *3.2.1 Moderate-Intensity Interval Training*

A specific interval training model is popular in the Central European cardiac rehabilitation centres that utilize a moderate-intensity interval approach, known as the "4/2" or "4/3" protocol, with 4-min active blocks at moderate intensity, separated by 2- or 3-min recovery phases. Intensity of active phases for a patient with a good functional capacity of 50–70% PWR and passive recovery at 0 Watts have been applied. For more deconditioned individuals' active phases, the intensity of 50% PWR and 3 min passive recovery phases at 0 watts have been suggested.

A graphical display of such training is presented in Figs. 3.6 and 3.7.

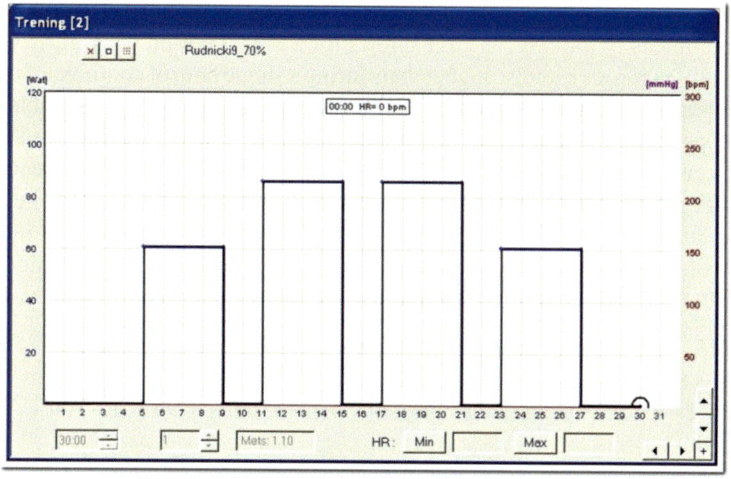


Fig. 3.6 Moderate-intensity interval training (MIIT 4/2) model, *source* by the authors

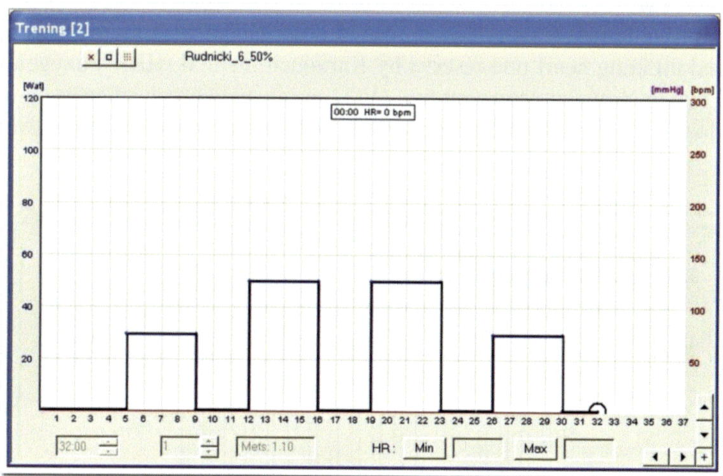


Fig. 3.7 Moderate-intensity interval training (MIIT 4/3) model, *source* by the authors

### *3.2.2 High-Intensity Interval Training*

High-intensity interval training (HIIT), the popularity of which is rapidly growing in cardiac rehabilitation, is a well-established training method for middle- and long-distance runners. The multiple golden medallists of the Olympic Games, Paavo Nurmi and Emil Zatopek, have been performing HIIT on a regular basis. HIIT consists of phases of intensive aerobic exercise with periods of passive or active recovery. Significant physiological differences prevail between HIIT and continuous moderate-intensity training, as HIIT allows for a longer period of exercise with the total time of vigorous exercise higher than during a single bout of continuous exercise. HIIT imposes a stronger cardiovascular stimulus that would not be sustainable during continuous mode [4]. Furthermore, patients find this form of training attractive, as the protocol is more diverse than it is during a constant workload due to the presence of recovery phases. HIIT allows for multiple combinations of its variables, i.e., active interval intensity and duration, recovery intensity and duration, exercise modality, number of repetitions, number of series, and between-series recovery duration.

Due to the duration of the active phases, HIIT has been divided into classes:

- s-HIIT- short-term cycles <60 s (typically 15–30 s) at 100% to 120% $VO_{2peak}$ (Fig. 3.8).
- m-HIIT—cycles of moderate duration of 1–3 min at 95% to 100% $VO_{2peak}$ (Fig. 3.9).
- l-HIIT: long duration cycles of 3 to 15 min at 85% to 90% $VO_{2peak}$ (Fig. 3.10).

HIIT typically incorporates 4–20 load cycles at high-to-maximal intensity, lasting from 10 s to 4 min, separated by regeneration phases at light or moderate intensity—active recovery (AR) or with very low or no load—passive recovery (PR). The recovery phase duration ranges from just 15 s up to 4 min. HIIT intensity may be prescribed utilizing heart rate reserve by Karvonen method (HRR), oxygen uptake reserve, maximal oxygen consumption (VO2 max), and perceived exertion by RPE Borg. Most HIIT prescription models are based on the relation between work and

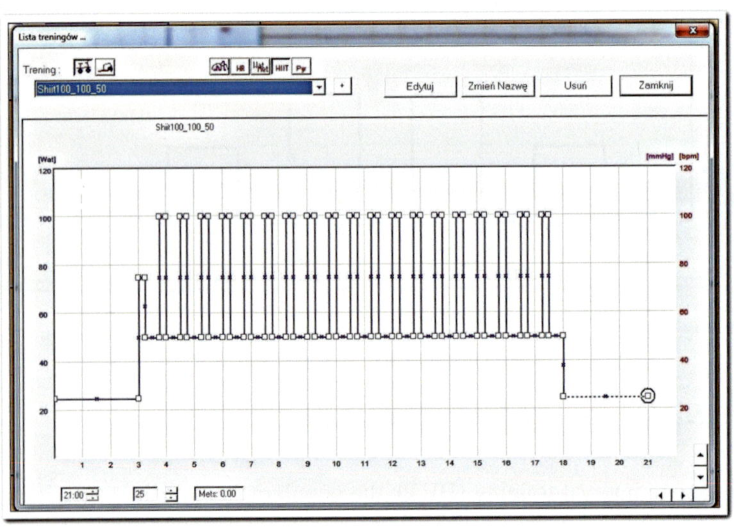


**Fig. 3.8** Short high-intensity interval training (s-HIIT) on the leg ergometer: 100% PWR (100 watts), time: 15 s, AR—50% PWR (50 watts) for 30 s, warm-up, and cool-down at 25% of PWR (25 watts) for 3 min (with permission from reference 2)

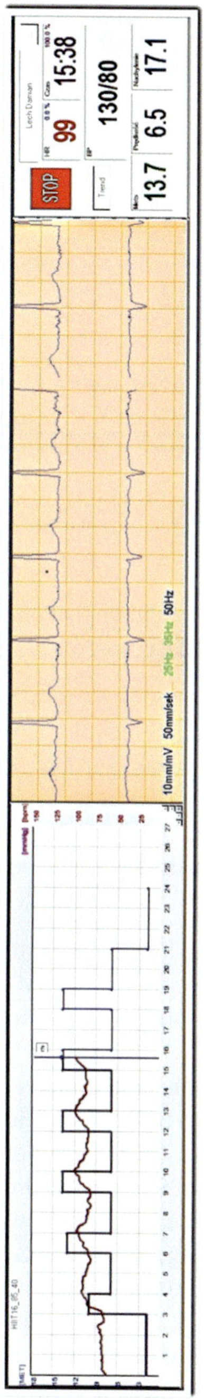


**Fig. 3.9** Monitored parameters: heart rate, load, blood pressure and time during moderate duration high intensity interval training (with permission from reference 2)

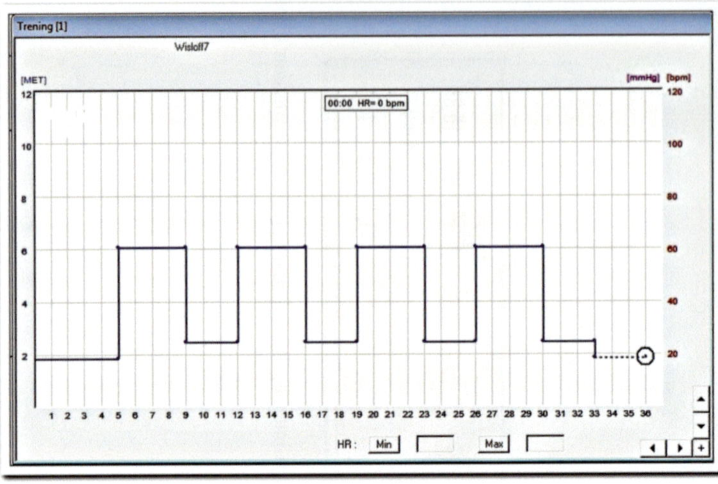


**Fig. 3.10** Long high intensity interval training (l-HIIT) on a treadmill: 85% of MET-R (peak MET = 7 MET), 4 × 4 min, AR—25% of MET-R (3 × 3 min) (with permission from reference 2)

recovery phases, and the latter has a powerful impact on performance, as documented in a study by Guiraud. Four different HIIT protocols for patients with coronary artery disease have been compared.

- 15 s of active phase at 100% of maximal aerobic power/15 s of passive recovery
- 15 s of active phase at 100% of maximal aerobic power/15 s of active recovery (50% of maximal aerobic power)
- 1 min of active phase at 100% of maximal aerobic power/1 min passive recovery
- 1 min of active phase at 100% of maximal aerobic power/1 min of active recovery (50% of maximal aerobic power).

The first model has documented the longest time to exhaustion, i.e., short (15 s) bouts of high-intensity load with a passive recovery emerged as the most efficacious [5]. Therefore, passive recovery allows for the more productive utilization of energetic substrates.

The example of m-HIIT has been illustrated in Fig. 3.9 [6]. 37-year-old patient after a recent myocardial infarction with ST-segment elevation (STEMI), who underwent percutaneous coronary angioplasty (PCI) on β-blocker therapy. Exercise testing data: resting HR of 50 bpm; peak HR of 98 bpm; resting BP of 130/80 mmHg; peak BP of 170/90 mmHg; 13.1 MET attained, i.e.,105% of predicted value; test terminated with a fatigue equal 17 RPE Borg; ECG showed ST depression by 0.15 mV. Exercise training model during the 3rd week of training: active phases–106% MET-R (13.7 MET), 60 s; AR—50% MET-R (7 MET), 120 s; warm-up and cool-down: intensity equal 2 MET for 3 min Source: by the authors.

The long-HIIT (l-HIIT) example has been presented in Fig. 3.10.

The most popular long HIIT model starts with a 10-min warm-up, followed by four hard segments of 4 min each at an intensity above the second lactate threshold (typically at 90% of peak HR), separated by 3-min recovery segments (at an intensity below the first lactate threshold in case of passive recovery or beyond the first lactate threshold in case of active recovery). Training ends with a few-minute cool-down phase [7, 8].

Training by Meyer is based on preceding steep ramp testing on a leg ergometer with calculated so-called maximal short-term exercise capacity (MSEC). HIIT progression occurs by shortening active phases with a simultaneous increase of intensity (up to 80% of the MSEC). The initial training incorporates 30-s active phases at 50% of MSEC and 60-s passive recovery (at 10 watts). Further progression includes gradual shortening of active phases duration up to 10 s with the escalation of training intensity up to 80% of the MSEC without a change in the recovery period [9].

Gayda suggested two ways of HIIT progression based on initial functional capacity level, gradually transitioning from s-HIIT to l-HIIT [10].

HIIT, as the final stage of the graded approach, has been proposed by the European Association of Preventive Cardiology. Somewhat different concepts of staged training evolving into HIIT are conducted at Mayo Clinic.

The European Association of Preventive Cardiology recommends prescribing exercise training concerning the patient's risk [2]. Low-risk patients, i.e., those with an uncomplicated course of cardiovascular events, clinically stable picture, good left ventricular performance, and functional capacity of >50% of the predicted value, commence an exercise training program with light-to-moderate-intensity continuous protocol, with the subsequent shift into the high-intensity interval approach.

Opposite, high-risk patients, i.e., those with dyspnea at NYHA class II-III, presence of arrhythmias, frailty, with functional capacity <50% of the predicted value (optimally evaluated from the cardiopulmonary exercise testing) should commence their exercise regimen from the low-intensity interval training, followed by the moderate-intensity continuous training. Selected high-risk patients may be suitable for a final transition into high-intensity interval training. The initial stage, i.e., low-intensity interval mode on a leg cycle ergometer, allows for instant and precise load changes, with active and recovery blocks of 30- and 60-s duration, respectively. After a few weeks of well-tolerated training progression, a continuous workload can be implemented at an intensity set between ventilatory thresholds or between 50 and 80% of the heart rate reserve.

The combined exercise training program conducted at Mayo Clinic starts with two weeks of moderate-intensity continuous training at RPE 11–14 to assess response to exercise and improve exercise tolerance [1]. The next stage (weeks 3–4) includes s-HIIT blocks at RPE 16–18, separated by 2-min active recovery phases. This, in turn, is changed into m-HIIT at the same RPE (weeks 5–8). The final maintenance stage includes l-HIIT, separated by 3-min active recovery phases. The summary of popular HIIT models is depicted in Table 3.1.

The study by Wislof has demonstrated reverse left ventricular remodeling after HIIT [11]. Subsequent phases of high-intensity exercise interspersed with

**Table 3.1** The high intensity interval training models with progression principle according to reference, by the authors

| HIIT model | Initial pattern | Progression | Final pattern |
|---|---|---|---|
| By Meyer [9] | 30 s active/60 s passive blocks at 50% of maximal intensity from ramp bike test | Increase of intensity Shortening of active blocks | 10 s active/60 s passive blocks at 80% of maximal intensity from ramp bike test |
| By Gayda (patients < 5 MET) [10] | s-HIIT 15 s active/15 s passive blocks | Lengthening of active blocks duration | m-HIIT 1-3 min active/1–3 min passive blocks |
| By Gayda (patients > 5 MET) [10] | s-HIIT 15-30 s active/ 15-30 s passive blocks | Lengthening of active blocks duration | l-HIIT 4 min active/ 3 min passive blocks |
| EAPC (low-risk patients) [2] | MICT at intensity between 1st and 2nd VT | MICT converted into HIIT | HIIT with active blocks at intensity > 2nd VT |
| EAPC (high-risk patients) [2] | LIIT with 30 s active blocks at intensity of 50% of PWR and 60 s passive blocks | LIIT converted into MICT | MICT 20–30 min at intensity between 1st and 2nd VT |
| By Mayo Clinic [1] | MICT 20–30 min at RPE 11–14 | MICT converted into HIIT | l-HIIT at RPE 15–18 |

EAPC, European Association of Preventive Cardiology; HIIT, high-intensity interval training; l-HIIT, long HIIT; LIIT, low-intensity interval training; m-HIIT moderate HIIT; MET, metabolic equivalents of task; MICT, moderate-intensity continuous training; RPE, rating of perceived exertion; s, second; s-HIIT, short HIIT

recovery phases may favorably affect the myocardium, mimicking the ischemia-preconditioning phenomenon and thus delivering constant cardioprotection [12].

Recent studies performed in animal models have demonstrated that intermittent ischemia provoked by HIIT results in the formation of collateral coronary vessels. HIIT, therefore, can be an attractive and efficacious alternative for patients with coronary artery disease or heart failure, which has been confirmed by two large multicenter studies comparing HIIT versus MICE (the SAINTEX-CAD and the SMARTEX-HF studies). The former encompassed 200 patients with coronary artery disease and preserved the left ventricular ejection fraction, whereas the latter included over 200 patients with heart failure with reduced left ventricular ejection fraction). Both studies demonstrated no inferiority of HIIT compared to the moderate-intensity continuous approach concerning oxygen uptake improvement [13, 14].

### 3.2.3 *Pyramidal Training*

Pyramids-shaped training has gained popularity since Tschentscher proposed such a training model consisting of three 8-min cycles and a load range of 65–95% of peak heart rate, separated by two active recovery phases at the intensity of 60–70% peak heart rate [15]. Training starts with a 5-min warm-up phase and ends with a 5-min cool-down phase, each at the intensity of 60–70% peak heart rate. Thus, pyramidal training encompasses low-to-moderate, moderate-to-vigorous, and vigorous-to-maximal intensities. An example of modified pyramidal training on a treadmill is exhibited in Fig. 3.11.

## 3.3 Customized Interval-to-Continuous Training

Interval-to-continuous exercise training (ITC) is based on a progressive transition from moderate-intensity interval training (MIIT) into an intermediate protocol between MIIT and MICT [16].

The detailed ITC characteristics include:

- frequency 5 per week for residential programs, 2–3 per week for outpatient
- training intensity on cycle ergometer or treadmill: moderate or moderate-to-vigorous (50–70% PWR, 50–80% HRR, RPE 12 to 15) for active phases and 0 Watts (ergometer) or 2 MET (V—2 km/h, inclination—0) for passive recovery phases.
- training duration 30–40 min, including 3–5 min of warm-up, active phases of 4–20 min, recovery phases of 2 min, and 3–5 min of cool-down phase
- training progression depends on the declared RPE value, chronotropic response for a given load, clinical picture, and risk level and is typically implemented through the increase in the number of load cycles, followed by the prolongation of active phase duration.

Optimally, the heart rate trajectory should be flat during active phases, with a symmetrical decrease by at least 10 beats per minute at the end of the recovery phase compared with that at the end of the active phase.

Typically, the initial exercise protocol includes 4-min active phases and 2-min passive recovery ("4/2" model) or 2-min active phases and 2-min passive recovery ("2/2" model) for patients with poor exercise capacity. The gradual progression incorporates a higher number of active phases (up to 30-40 min of total training duration), followed by a prolongation of active blocks duration (by 2 min, i.e., from "2/2" into "4/2" model or from" 4/2″ into "6/2″ model, with concomitant shortening of training total time.

Patients with initial poor exercise capacity attain at the end of the exercise training program the protocol of "6/2" or "8/2", whereas the final exercise protocol of those with better functional capacity is "15/2" or "20/2" (Fig. 3.12).

ITC appears to be motivating, safe and efficacious exercise training protocol with abovementioned progression principle, whereas maintaining determined training heart range. It can be offered as a follow-up training (walking outdoors, stationary biking, Nordic Walking, jogging) offered to patients on discharge.

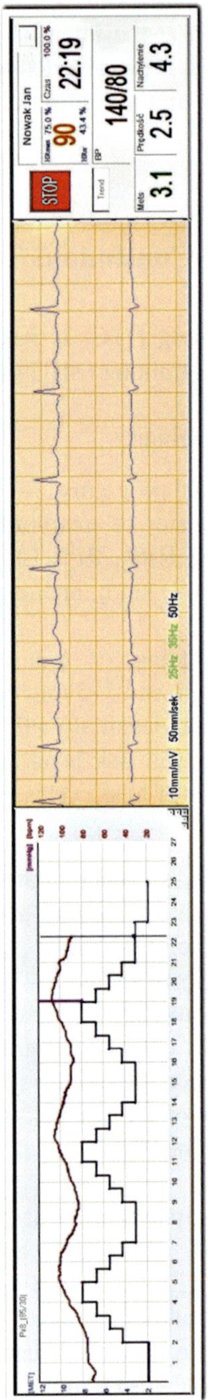


**Fig. 3.11** Pyramidal training on treadmill (with permission from reference 2)

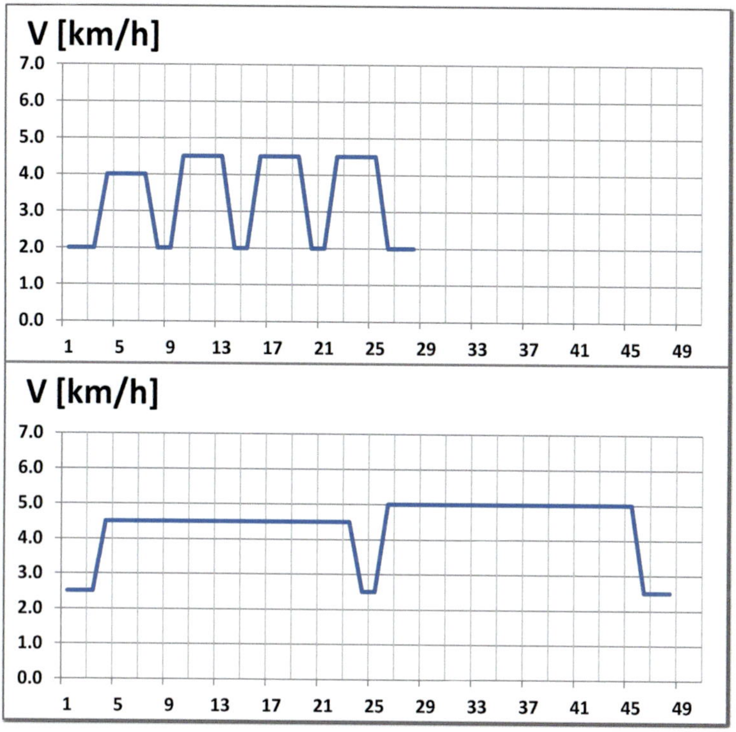


**Fig. 3.12** An example of the initial and final ITC training protocol, *Source* by the authors, ITC, interval-to -continuous

The key features of continuous, interval, and interval-to-continuous approaches have been summarized in Fig. 3.13.

## 3.4 Resistance Training

Resistance training (RT) combines isokinetic exertion (by generating force at a fixed angular velocity) and isotonic effort (by generating force when changing the length of the muscle). It aims to improve muscle strength, which decreases naturally with age or because of a sedentary lifestyle or prolonged immobilization (15). Beneficial effects of resistance exercise include:

- increase of muscle mass and strength -muscle endurance
- lowering the blood pressure values and the double product value
- decreased insulin resistance
- improvement of the lipid profile
- preventing the loss of bone mineral density

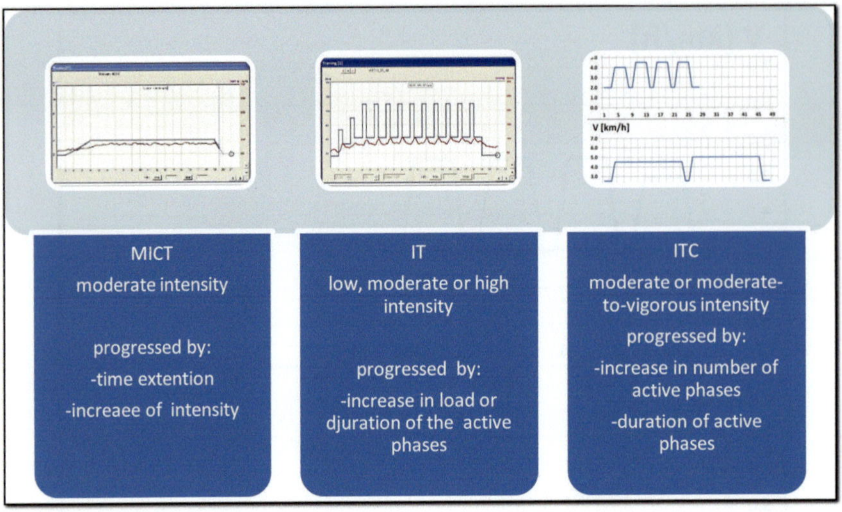


**Fig. 3.13** Key features of the continuous, interval and intermediate training protocols, *Source* by the authors IT, interval training, ITC, interval-to-continuous training, MICT, moderate intensity continuous training

- relief of the pain in degenerative joint diseases
- improved balance (reduced the risk of falls)

Absolute contraindications to resistance training include:

- unstable angina,
- decompensated heart failure,
- uncontrolled arrhythmias,
- uncontrolled hypertension (>180/100 mmHg),
- severe symptomatic aortic stenosis,
- severe pulmonary hypertension,
- acute myo-, peri- and endocarditis,
- aortic dissection,
- Marfan disease,
- active proliferative retinopathy (for high-intensity resistance training).

The relative contraindications entail:

- uncontrolled hypertension > 160/100 mmHg,
- low initial functional capacity < 4 MET.

Resistance training sessions should be terminated in case of:

- chest pain,
- significant fatigue,
- dizziness,

- abnormal chronotropic or pressure response (decrease in heart rate, lack of increase or decrease in blood pressure, with concomitant symptoms like angina or dyspnea,
- SBP > 200 mmHg and/or DBP > 110 mmHg).

Resistance training can be conducted with dedicated equipment (free weights, barbells, dumbbells, medicine balls, weight machines, elastic bands) or utilizing body resistance (e.g., squats).

It is recommended that the upper body receive an initial intensity of 30–40% of 1-RM (one-repetition maximum) and the lower body receive -50–60% of 1-RM [17].

Resistance training should be conducted 2–3 times per week, lasting at least 48 h between sessions.

It should involve each major muscle group, target agonists, and antagonists, and can be planned as an independent session or as a component of the warm-up or cool-down phases.

The general principle of resistance training has been listed in Fig. 3.14.

Training intensity is based on graded stress testing (with an estimated % of 1-RM), based on load–the repetition relationship as follows (Table 3.2).

The dedicated formulas for 1-RM estimation from multiple RM have been suggested, as follows [18]:

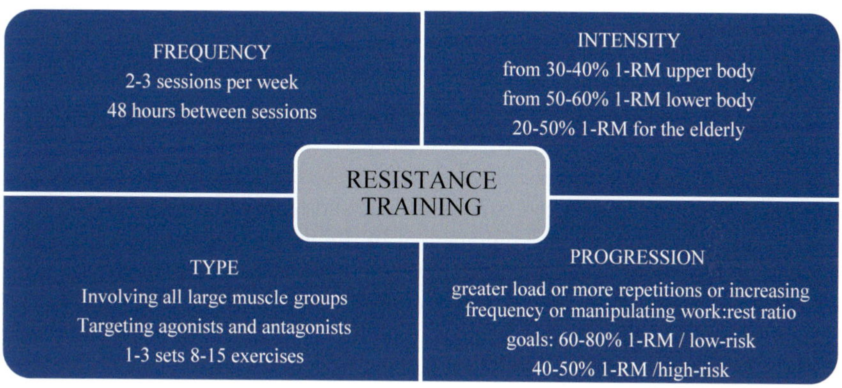


**Fig. 3.14** Resistance training principle, Abbreviations: 1-RM -one- repetition maximum. *Source* by the authors

**Table 3.2** The load-repetition relationship

| Percentage of 1-RM | Related number of repetitions |
|---|---|
| 60 | 17 |
| 70 | 12 |
| 80 | 8 |
| 90 | 5 |
| 100 | 1 |

$$1RM = (1 + 0{,}0333 \times \mathit{repetitions}) \times \mathit{applied\ weight}$$

where: 1RM, one repetition maximum.

Moreover, a dedicated scale (OMNI-RES) has been established to facilitate the progression of resistance training. It supports tracking the perceived intensity during strength training utilizing visual, numerical, and verbal perceptual exercise intensity equivalents from 0 points ("extremely easy") to 10 points ("extremely hard"). The usefulness of the OMNI-RES scale in the elderly population has been well documented [19].

It is also important to deliver sufficient instructions to patients prior to the commencement of strength training.

Such preliminary instruction should address the appropriate weight loads, safe lifting technique, range of motion, and breathing pattern (to avoid Valsalva maneuver). Thus, focusing on the correct breathing technique is of special importance. The lifting weights should be performed rhythmically at a moderate to slow speed through a full range of motion, breath-holding should be avoided by breathing out during the exertion phase, and the upper- and lower-body muscle groups should be alternated to allow for adequate rest.

An exercise specialist should supervise the training. In patients with recently implanted cardiac electrical devices, the cardiac rehabilitation physician should provide clearance to start resistance training to avoid lead fracture or dislodgment. Strength training of the upper body should be avoided for the first three months after valvular or coronary bypass surgery with full sternotomy to ensure full healing of the sternum.

To improve muscle strength, 8–12 repetitions are recommended, whereas 15 to 20 repetitions are needed to improve endurance.

Resistance training progresses through an increased number of repetitions, heavier load, and change in work-to-rest time ratio (i.e., shortening the rest period).

Resistance training can be implemented in different ways, e.g., for residential programs, typically, patients start with warm-up exercises in the morning, with added flexibility and mobility exercises. It is followed by the main aerobic training, and then after break, they complete strength training of 30 min to 1 h. For outpatient facilities, resistance training can be performed on alternative days or on the same day as the aerobic component (in this scenario, patients typically start with aerobic training first)—Fig. 3.15.

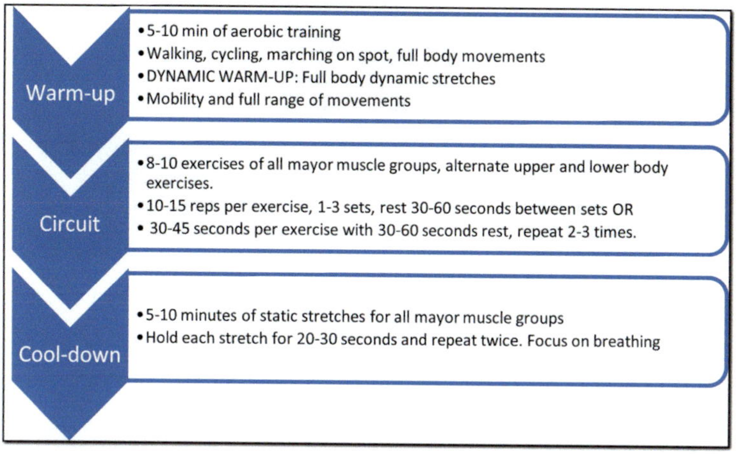


**Fig. 3.15** Components of a resistance training session, by the authors

Combining aerobic and resistance exercises improves peak aerobic fitness and lower and upper body strength.

A typical resistance training circuit encompasses chest press, shoulder press, triceps extension, biceps curl, pull-down (for upper back), lower back extension, abdominal crunches, quadriceps extension or leg press, and calf raise.

It is important to focus not only on strengthening the muscles through resistance training but also on balance and proprioceptive training, flexibility and mobility, exercises to strengthen the full body, and to improve patients' daily lives and activities by improving their functional and daily movements.

Full body stretches and mobility are presented from Figs. 3.16, 3.17, 3.18 and 3.19.

An example of a resistance training circuit for deconditioned patients is presented in Figs. 3.20, 3.21, 3.22, 3.23, 3.24, 3.25, 3.26, 3.27 and 3.28.

Figure 3.29 presents an example of the resistance training circuit for facilities with minimal equipment or performed as home training, with pictures from Figs. 3.30, 3.31, 3.32, 3.33, 3.34, 3.35, 3.36 and 3.37.

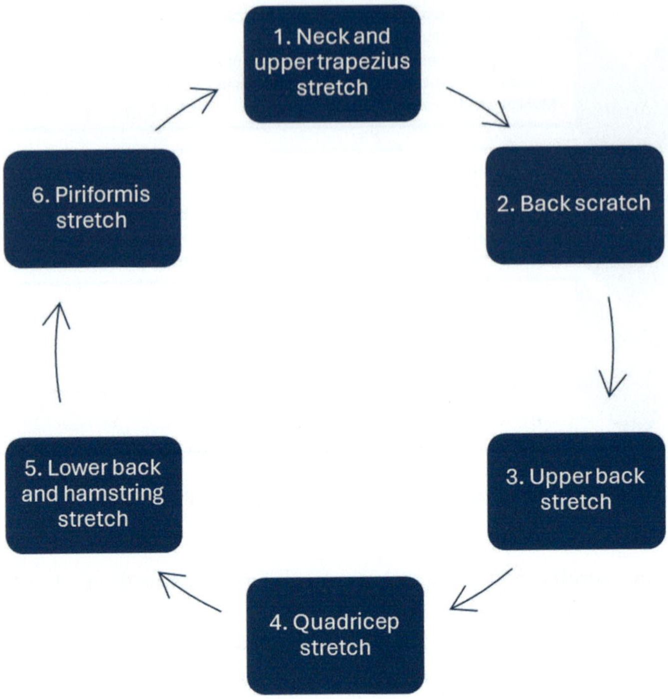


**Fig. 3.16** Example of stretching exercises. Provided by authors

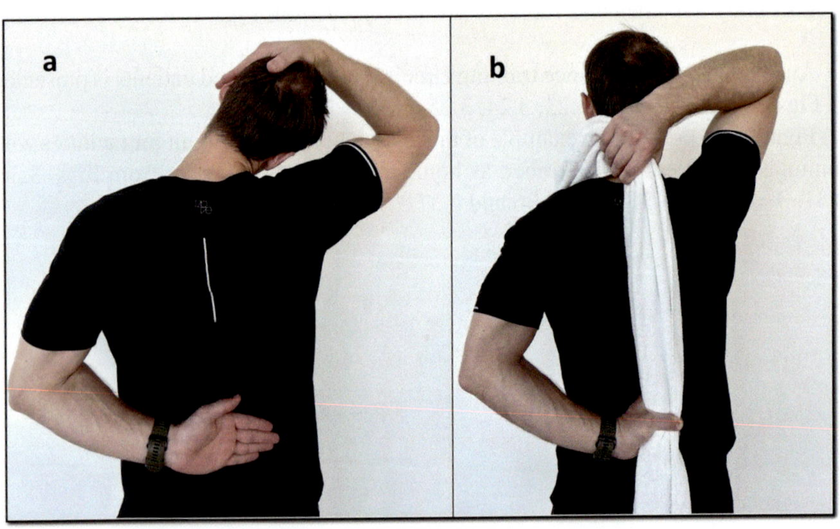


**Fig. 3.17** **a** Neck and upper trapezius stretch, **b** Back scratch (shoulder mobility)

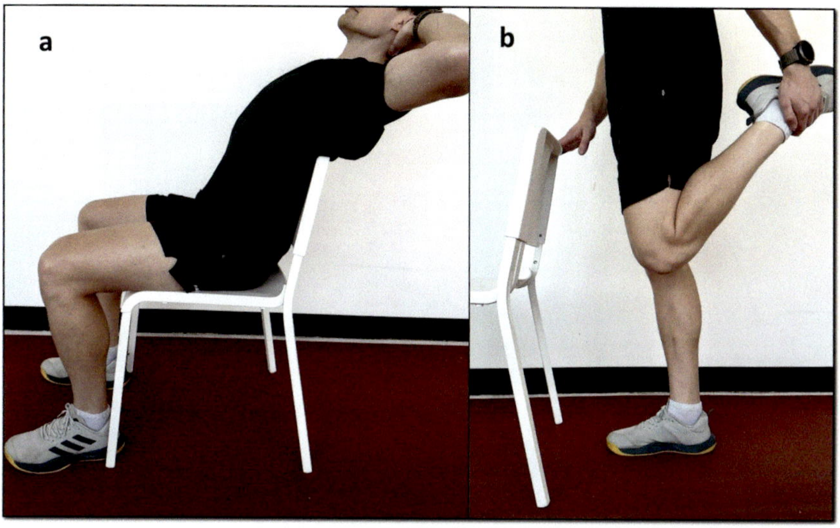


**Fig. 3.18** **a** Upper back stretch (thoracic spine mobility), **b** Quadriceps stretch

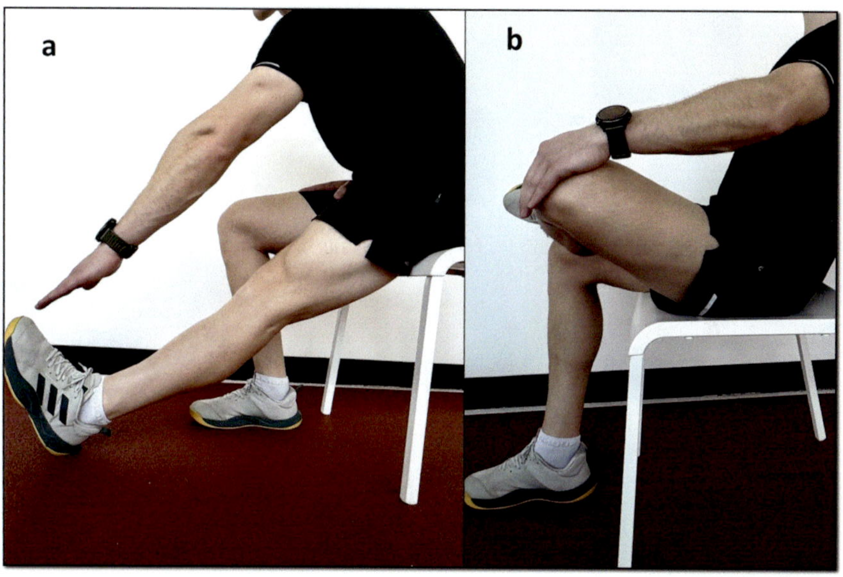


**Fig. 3.19** **a** Lower back and hamstring stretch, **b** Piriformis stretch

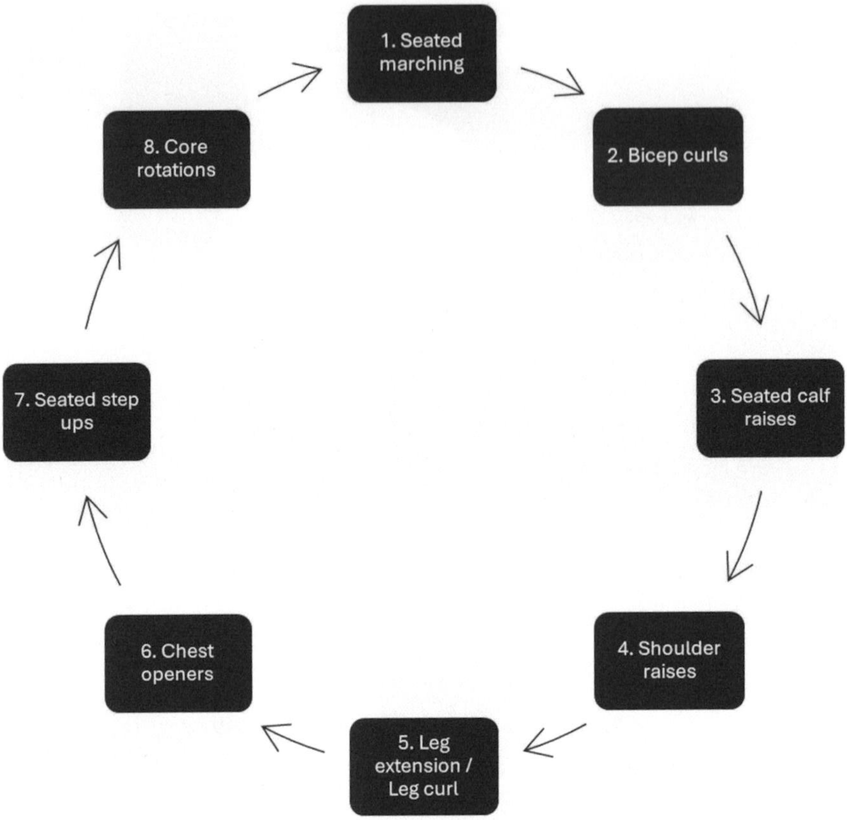


**Fig. 3.20** Resistance training- circuit for very weak patients

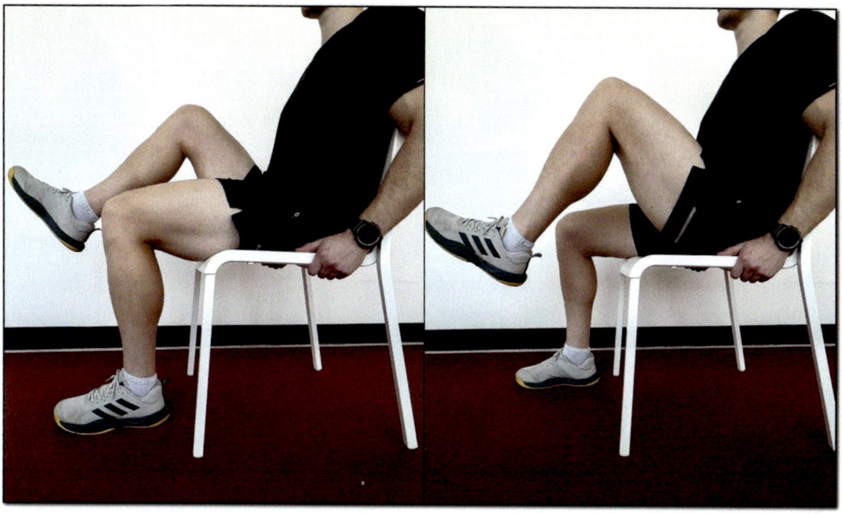

**Fig. 3.21** Seated marching

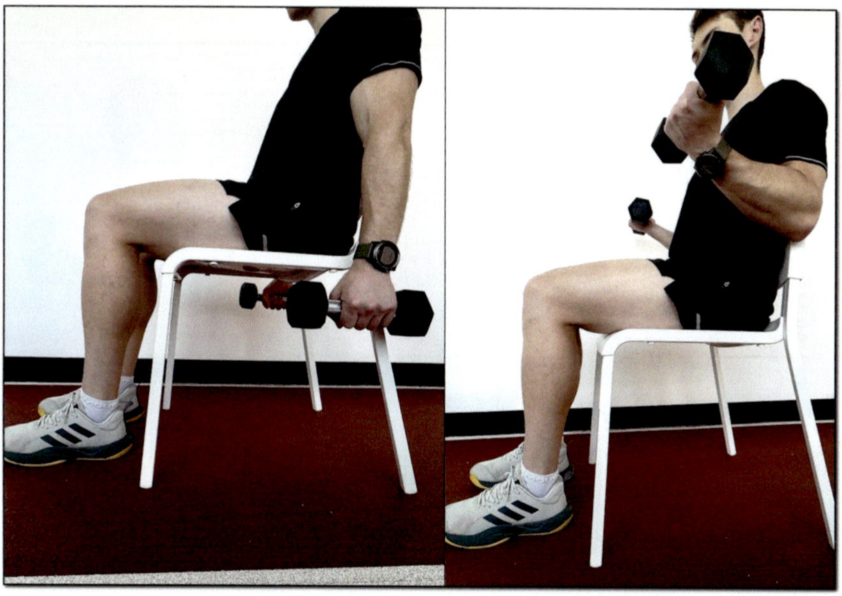

**Fig. 3.22** Bicep curls

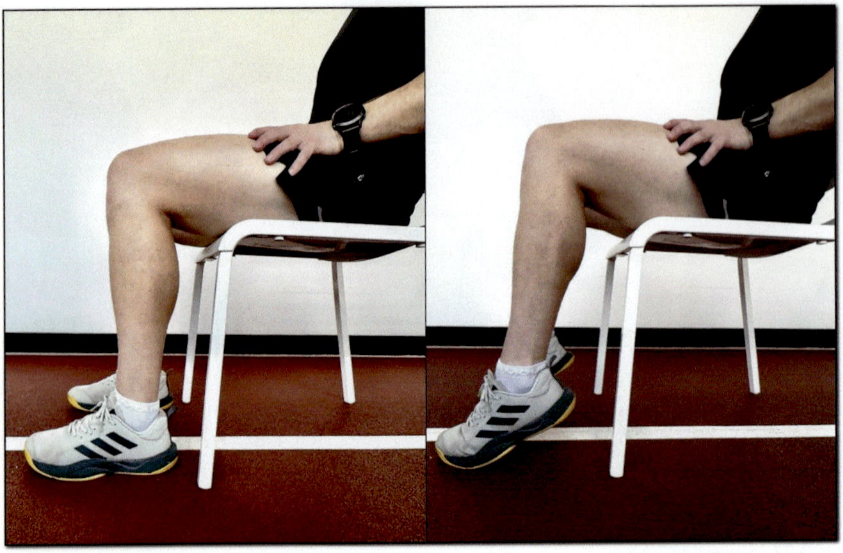

**Fig. 3.23** Seated calf raise

**Fig. 3.24** Shoulder raises

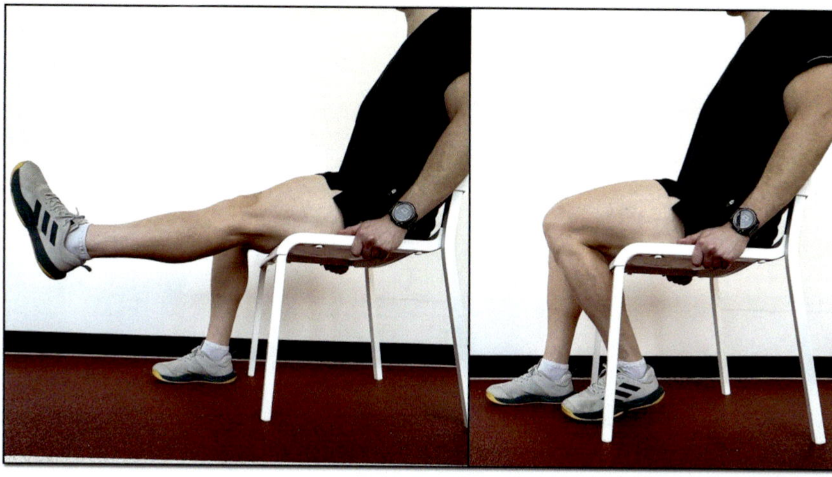

**Fig. 3.25** Seated leg extension/leg curl

**Fig. 3.26** Chest opener

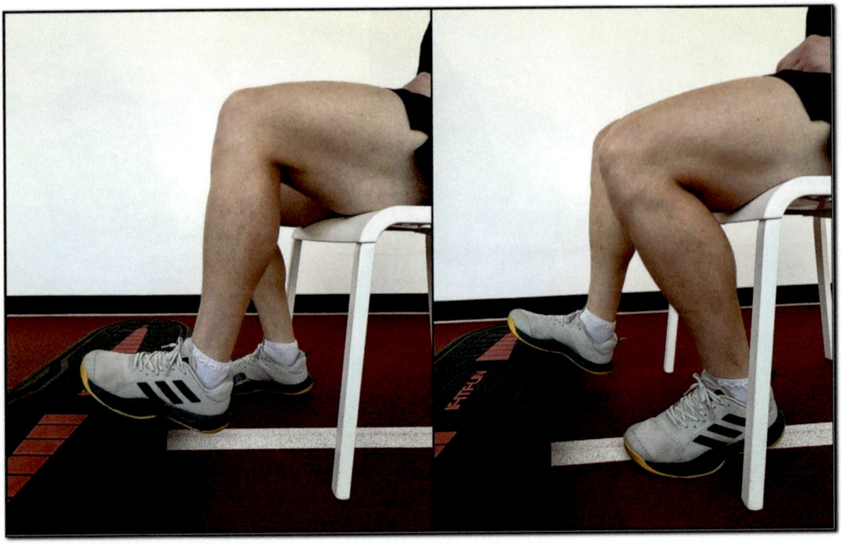

**Fig. 3.27** Seated step ups

**Fig. 3.28** Seated core rotations

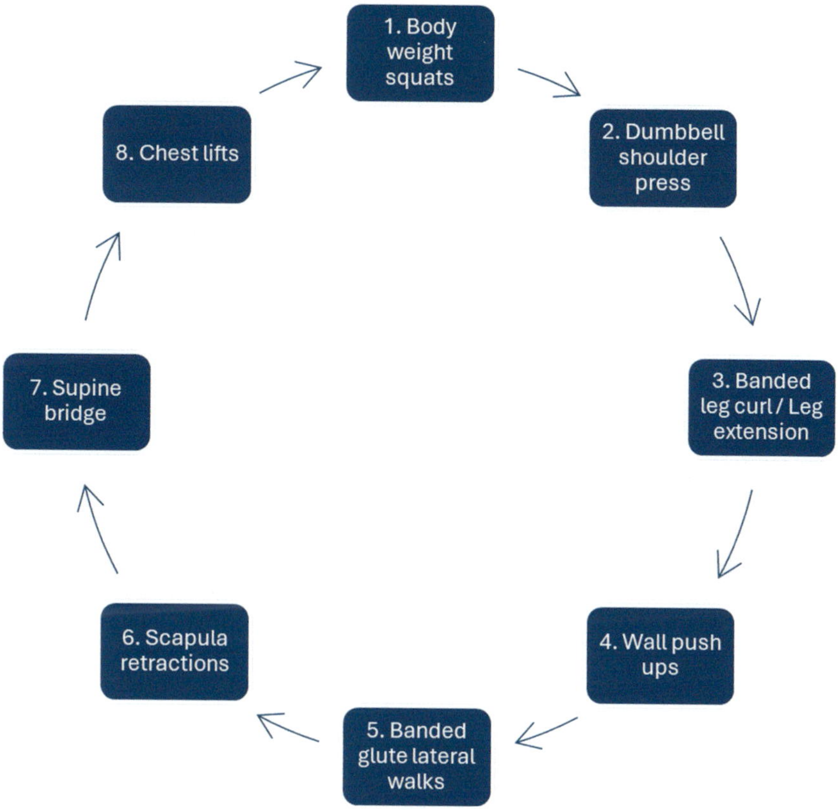


**Fig. 3.29** Resistance training-circuit with minimal equipment

Fig. 3.30 Body weight squats

Fig. 3.31 Dumbbell shoulder press

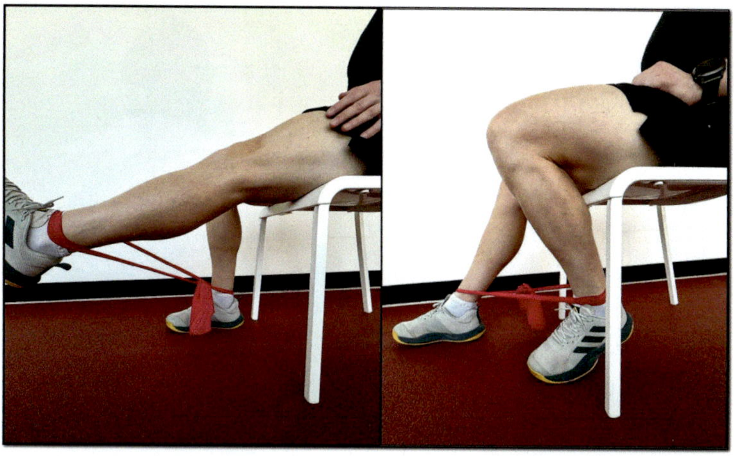

Fig. 3.32 Banded leg extension/leg curl

Fig. 3.33 Wall push up

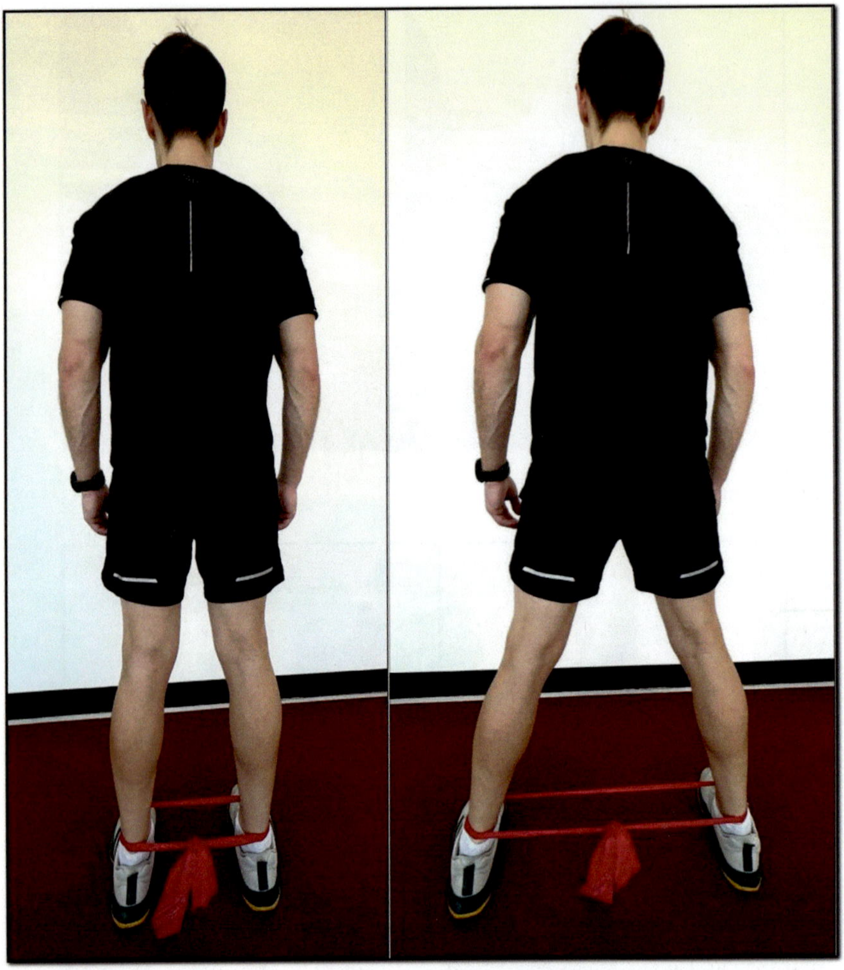

**Fig. 3.34** Banded glute lateral walks

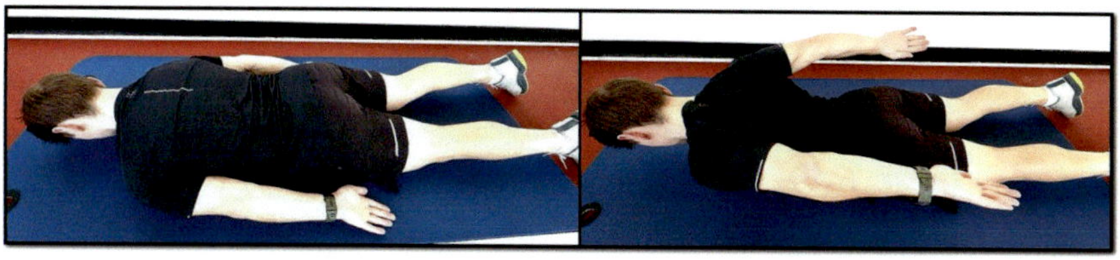

**Fig. 3.35** Prone scapular retractions

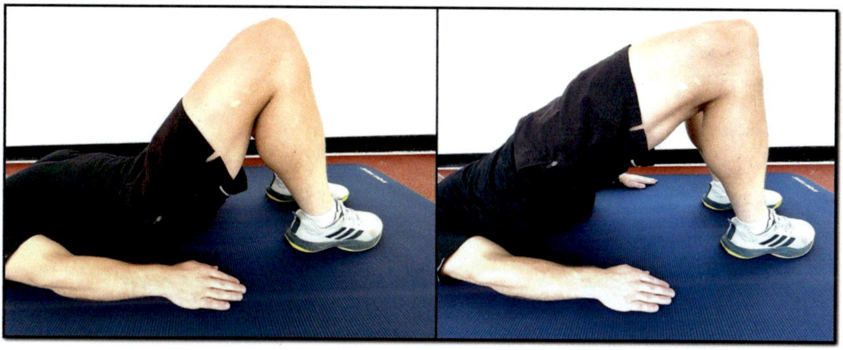

**Fig. 3.36** Supine bridge

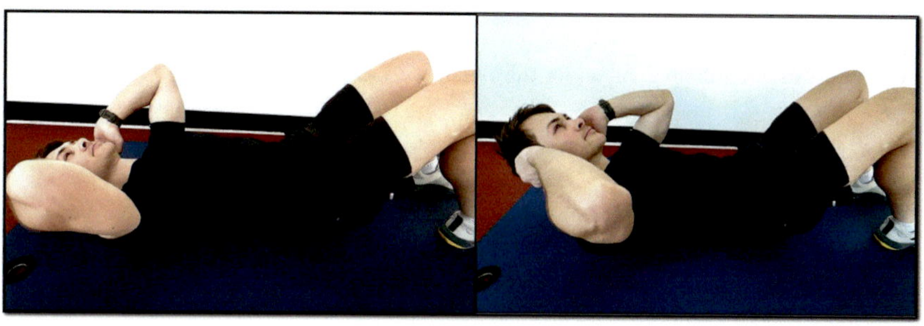

**Fig. 3.37** Chest lift

An example of a resistance training-circuit for fully equipped facilities is presented in Fig. 3.38 and in pictures (Figs. 3.39, 3.40, 3.41, 3.42, 3.43, 3.44, 3.45 and 3.46).

The balance training exercises are presented in Figs. 3.47 and 3.48.

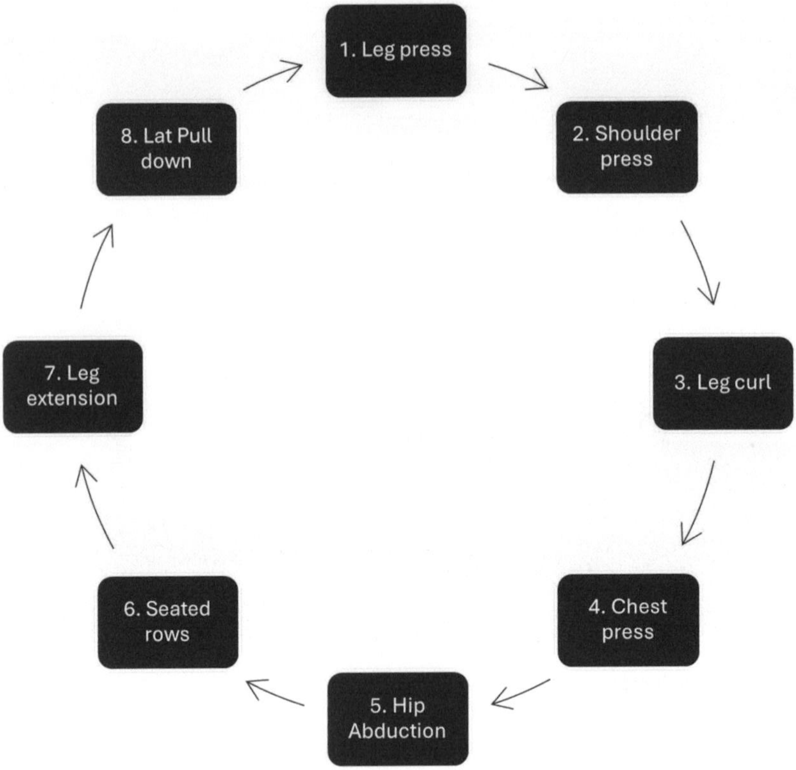


**Fig. 3.38** Resistance training-circuit for fully equipped facility

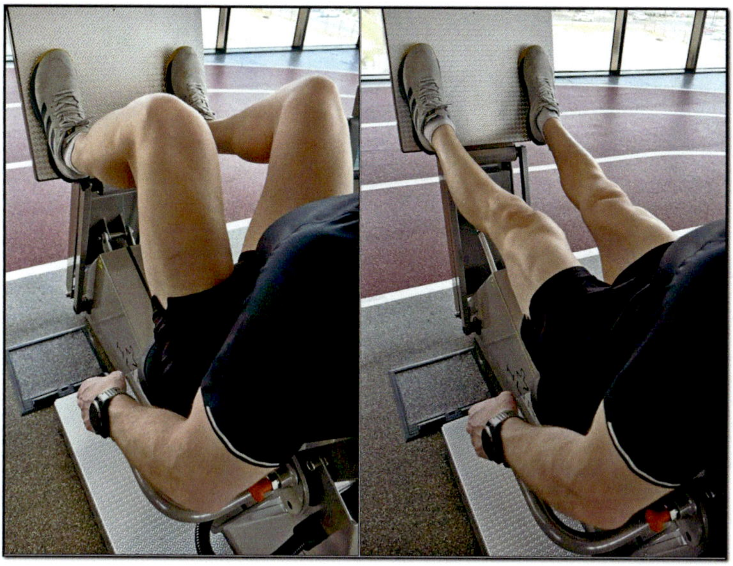

Fig. 3.39 Leg press

Fig. 3.40 Shoulder press

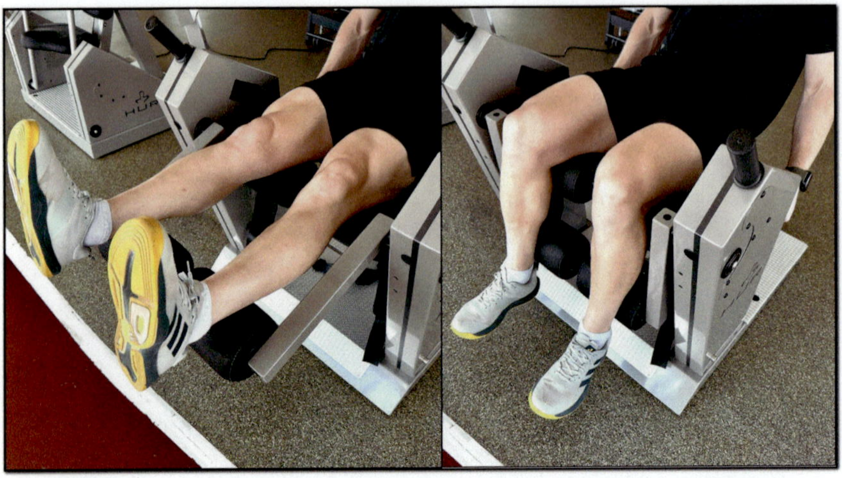

**Fig. 3.41** Leg curl

**Fig. 3.42** Chest press

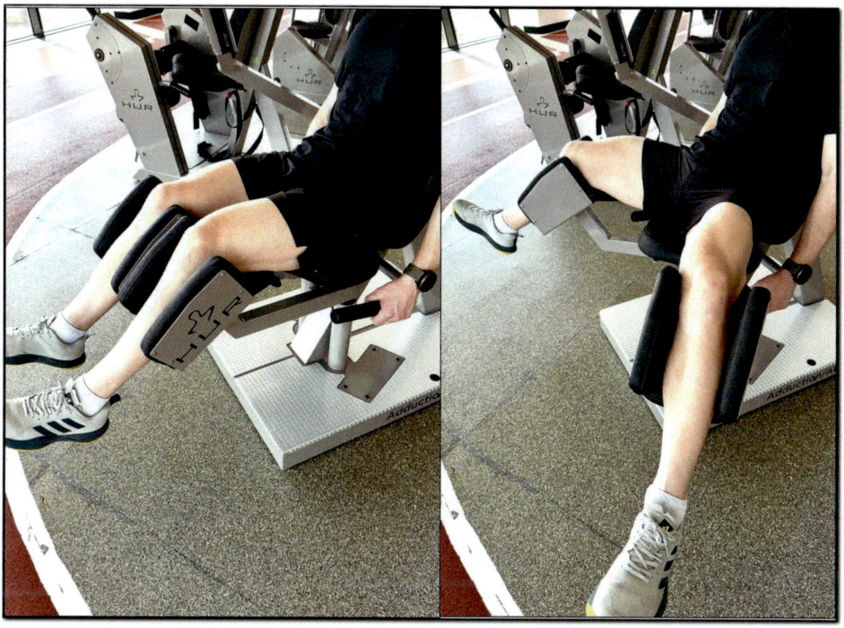

Fig. 3.43 Hip abduction

Fig. 3.44 Seated row

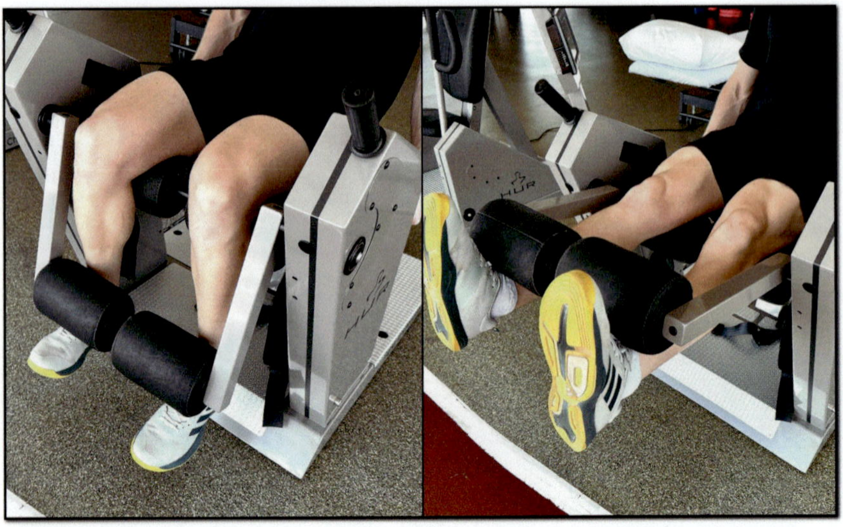

Fig. 3.45 Leg extension

Fig. 3.46 Lat pull down

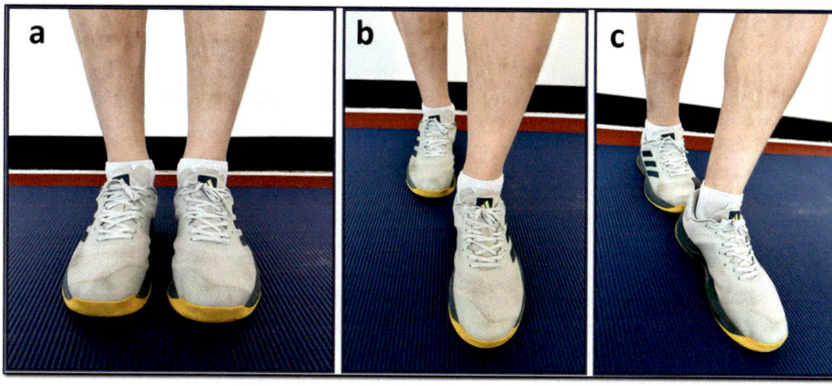


**Fig. 3.47** **a** Level 1: Romberg stance, **b** Level 2: tandem stance 1, **c** Level 3: tandem stance 2

**Fig. 3.48** Level 4: single leg stance/unipedal stance

**AuthorContribution** **Competing Interests** The authors have no conflicts of interest to declare that are relevant to the content of this chapter.

## References


1. Taylor JL, Bonikowske AR, Olson TP. Optimizing outcomes in cardiac rehabilitation: the importance of exercise intensity. Front Cardiovasc Med. 2021;8: 734278. https://doi.org/10.3389/fcvm.2021.734278.
2. Abreu A, Schmid JP, Piepoli MF. The ESC handbook of cardiovascular rehabilitation. Oxford, UK: Oxford University Press; 2022.
3. Vanhees L, Rauch B, Piepoli M, van Buuren F, Takken T, Börjesson M et al.; Writing Group, EACPR. Importance of characteristics and modalities of physical activity and exercise in the management of cardiovascular health in individuals with cardiovascular disease (Part III). Eur J Prev Cardiol. 2012;19:1333–56. https://doi.org/10.1177/2047487312437063.
4. Mezzani A, Hamm LF, Jones AM, McBride PE, Moholdt T, Stone JA, et al. European Association for Cardiovascular Prevention and Rehabilitation; American Association of Cardiovascular and Pulmonary Rehabilitation; Canadian Association of Cardiac Rehabilitation. Aerobic exercise intensity assessment and prescription in cardiac rehabilitation: a joint position statement of the European Association for Cardiovascular Prevention and Rehabilitation, the American Association of Cardiovascular and Pulmonary Rehabilitation and the Canadian Association of Cardiac Rehabilitation. Eur J Prev Cardiol. 2013;20:442–67. https://doi.org/10.1177/2047487312460484.
5. Pattyn N, Beulque R, Cornelissen V. Aerobic interval vs. continuous training in patients with coronary artery disease or heart failure: an updated systematic review and meta-analysis with a focus on secondary outcomes. Sports Med. 2018:1189–1205. https://doi.org/10.1007/s40279-018-0885-5.
6. Guiraud T, Nigam A, Gremeaux V, Meyer P, Juneau M, Bosquet L. High-intensity interval training in cardiac rehabilitation. Sports Med. 2012;42:587–605. https://doi.org/10.2165/11631910-000000000-00000. PMID: 22694349.
7. Helgerud J, Høydal K, Wang E, Karlsen T, Berg P, Bjerkaas M, et al. Aerobic high-intensity intervals improve VO2max more than moderate training. Med Sci Sports Exerc. 2007;39:665–71. https://doi.org/10.1249/mss.0b013e3180304570.
8. Bækkerud FH, Solberg F, Leinan IM, Wisløff U, Karlsen T, Rognmo Ø. Comparison of three popular exercise modalities on V˙O2max in Overweight and obese. Med Sci Sports Exerc. 2016;48:491–8. https://doi.org/10.1249/MSS.0000000000000777.
9. Meyer K, Foster C, Georgakopoulos N, Hajric R, Westbrook S, Ellestad A, et al. Comparison of left ventricular function during interval versus steady-state exercise training in patients with chronic congestive heart failure. Am J Cardiol. 1998;82:1382–7. https://doi.org/10.1016/s0002-9149(98)00646-8.
10. Ribeiro PAB, Boidin M, Juneau M, Nigam A, Gayda M. High-intensity interval training in patients with coronary heart disease: prescription models and perspectives. Ann Phys Rehabil Med. 2017;60:50–7. https://doi.org/10.1016/j.rehab.2016.04.004.
11. Wisløff U, Støylen A, Loennechen JP, Bruvold M, Rognmo Ø, Haram PM, et al. Superior cardiovascular effect of aerobic interval training versus moderate continuous training in heart failure patients: a randomized study. Circulation. 2007;115:3086–94. https://doi.org/10.1161/CIRCULATIONAHA.106.675041.
12. Guiraud T, Nigam A, Juneau M, Meyer P, Gayda M, Bosquet L. Acute responses to high-intensity intermittent exercise in CHD patients. Med Sci Sports Exerc. 2011;43:211–7. https://doi.org/10.1249/MSS.0b013e3181ebc5de.
13. Ellingsen Ø, Halle M, Conraads V, Støylen A, Dalen H, Delagardelle C et al. SMARTEX heart failure study (study of myocardial recovery after exercise training in heart failure) group. High-Intensity interval training in patients with heart failure with reduced ejection fraction. Circulation. 2017;135:839–849. https://doi.org/10.1161/CIRCULATIONAHA.116.022924.
14. Conraads VM, Pattyn N, De Maeyer C, Beckers PJ, Coeckelberghs E, Cornelissen VA, et al. Aerobic interval training and continuous training equally improve aerobic exercise capacity in patients with coronary artery disease: the SAINTEX-CAD study. Int J Cardiol. 2015;179:203–10. https://doi.org/10.1016/j.ijcard.2014.10.155.

15. Tschentscher M, Eichinger J, Egger A, Droese S, Schönfelder M, Niebauer J. High-intensity interval training is not superior to other forms of endurance training during cardiac rehabilitation. Eur J Prev Cardiol. 2016;23:14–20. https://doi.org/10.1177/2047487314560100.
16. Staron A, Wolszakiewicz J, AlSulaimi M. Exercise prescription in cardiac rehabilitation. Basel: MDPI 2022.
17. Williams MA, Haskell WL, Ades PA, Amsterdam EA, Bittner V, Franklin BA et al. American heart association council on clinical cardiology; American heart association council on nutrition, physical activity, and metabolism. Resistance exercise in individuals with and without cardiovascular disease: 2007 update: a scientific statement from the American Heart Association council on clinical cardiology and council on nutrition, physical activity, and metabolism. Circulation. 2007;116:572–84.https://doi.org/10.1161/CIRCULATIONAHA.107.185214.
18. Brzycki M. Strength testing-prediction a one rep maximum from reps-to fatigue. J Phys Educ Recr Dance. 1993;64:88–90.
19. Gearhart RF Jr, Lagally KM, Riechman SE, Andrews RD, Robertson RJ. Safety of using the adult OMNI resistance exercise scale to determine 1-RM in older men and women. Percept Mot Skills. 2011;113:671–6. https://doi.org/10.2466/10.15.PMS.113.5.671-676.

# Chapter 4
# Exercise Prescription for Specific Populations

**Adam Staron and Amal Albatini**

**Abstract** Exercise-based cardiac rehabilitation has been traditionally linked to the post-myocardial infarct population. The benefits of regular exercise training, however, encompass not only patients with coronary artery disease. Contemporary cardiac rehabilitation evolved into a comprehensive and multi-factorial program covering the expanded population, including coronary artery disease, cardiac surgery, heart failure, implanted cardiac electrical devices, peripheral arterial disease, frailty, arrhythmia, and individuals with a high cardiovascular risk-factor profile. Delivery of exercise intervention should be personalized, based on specific diseases, comorbidities, clinical picture, and adaptation to imposed training load.

**Keywords** Heart failure · Coronary artery bypass surgery · Heart transplantation · Left ventricular assist device · Peripheral arterial disease · Cardiovascular risk factors

## 4.1 Coronary Artery Bypass Graft Surgery and Valve Surgery

Patients undergoing cardiac surgeries may suffer from post-operative symptoms, e.g., pain and fatigue and decline of physical functioning due to the deconditioning of the cardiovascular and muscular systems. All patients undergoing cardiac surgery should be enrolled in a cardiac rehabilitation program due to the favorable effects on reduction in all-cause mortality, independent of age, sex, prior myocardial infarction, or the presence of diabetes [1]. Patients undergoing cardiac surgeries may suffer from post-operative symptoms, e.g., pain and fatigue and decline of physical functioning due to the deconditioning of the cardiovascular and muscular systems. All patients

A. Staron (✉)
The Upper Silesian Medical Center Katowice Poland, Katowice, Poland
e-mail: astaron1973@wp.pl

A. Albatini
Ministry of Health Kuwait, Sulaibkhat, Kuwait

 
J. Rybicki and A. Staron (eds.), *Exercise Programming for Cardiac Patients*,
https://doi.org/10.1007/978-3-031-88475-7_4

undergoing cardiac surgery should be enrolled in a cardiac rehabilitation program due to the favorable effects on reduction in all-cause mortality, independent of age, sex, prior myocardial infarction, or the presence of diabetes [1].

Exercise training should be started in the early in-hospital phase and continued as a residential or outpatient program after discharge. Upper-body training can begin when chest stability is obtained, usually after six weeks. Exercise training programs should be individually tailored according to the clinical picture, exercise capacity on entry left ventricular function, and type of valve surgery [2]. After the cardiac surgery procedure, early mobilization is encouraged during hospitalization to avoid bed rest complications and accelerate recovery. Phase II of cardiac rehabilitation should optimally begin four weeks after surgery and may commence earlier in a center experienced with care after cardiac surgery. Patients undergoing coronary artery bypass graft procedure or valve surgery are typically older, with numerous comorbidities, and they have a lower functional capacity.

Before the commencement of the exercise training, a functional capacity assessment should be performed (the six-minute walk test (6MWT), symptom-limited exercise test (ExT), or submaximal ExT). The maximal ExT should not be executed within the four weeks following cardiac surgery. Exercise training evokes numerous cardiovascular and peripheral adaptations. The training intensity should reflect the current clinical outlook, left ventricular performance, arrhythmia or ischemia, functional capacity level, and surgery-related complications [3].

Valve surgery is typically performed at the advanced stage of heart failure; thus, improving functional capacity and/or the left ventricular systolic function occurs more slowly. The prevalence of frailty among patients referred to transcatheter aortic valve implantation is high, and an appropriate exercise training program should be applied [4]. The specific issues should be addressed after valve surgery (echocardiographic assessment of the transvalvular pressure gradient, valvular insufficiency, and presence of pericardial effusion). The Borg scale is a useful tool in the case of patients with atrial fibrillation. The anticoagulation regimen should be regularly monitored.

After median sternotomy, to avoid or limit sternal complications (the abnormal motion of the sternum), it is recommended to check regularly for sternum instability by assessing pain or discomfort and excessive sternal movement. If there is evidence of sternal instability, then lower extremities exercises should be performed only. Because of sternotomy, the range of motion (ROM) and exercises involving upper limbs should be modified 3 months after surgery. Instructions can be given to the patient when median sternotomy is as follows:

- restrictions in ROM and lifting weights more than 4 kg in the first 8–12 weeks,
- avoid laying or leaning on the side for 8–12 weeks,
- not performing any activities that require tensile stress, such as opening a sealed box.

The factors related to cardiac surgery that affect exercise intervention are presented in Fig. 4.1.

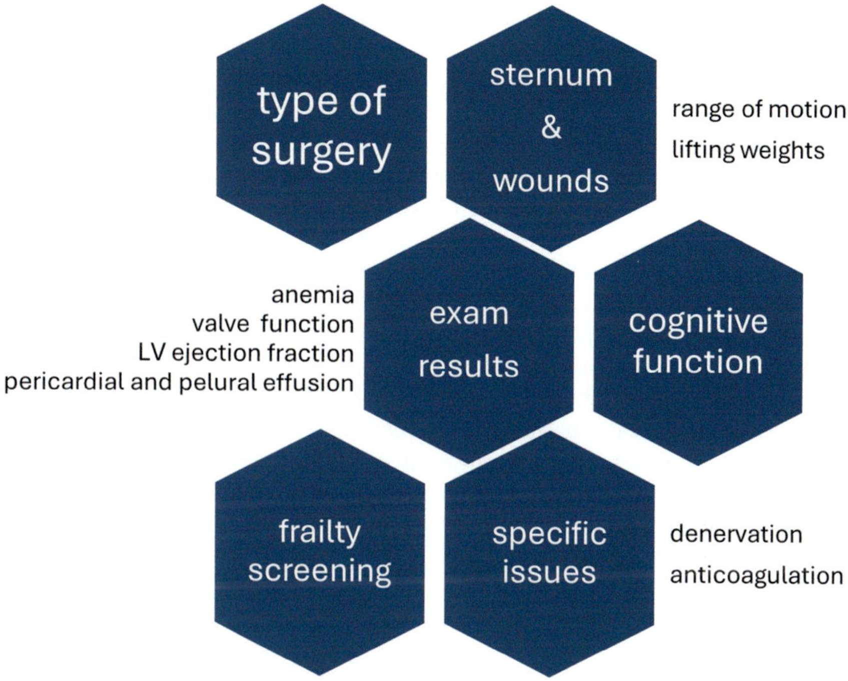


**Fig. 4.1** Cardiac surgery-related factors affecting cardiac rehabilitation process, by the authors LV, left ventricular

## 4.2 Atrial Fibrillation

It has been documented that atrial fibrillation (AF) incidence is lower in patients who exercise regularly at light or moderate intensity; however, exercising at vigorous intensity is associated with an increase in AF prevalence [5].

The unreliability of the heart rate (HR) should be considered when prescribing exercise. Exercise testing is useful for assessing myocardial ischemia, evaluating chronotropic response, and identifying signs and symptoms.

Exercise intensity should ideally be prescribed based on METs (training on the treadmill), work rate (bike training), and perceived exertion levels. The target session duration should be 60 min with adequate warm-up and cool-down.

## 4.3 Heart Failure

Chronic heart failure (CHR) develops because of structural and/or functional cardiac abnormality and results in insufficient cardiac output and/or elevated intracardiac pressures.

The current classification of heart failure primarily considers the left ventricular ejection fraction value [6]. It is defined as heart failure with reduced, mildly reduced, or preserved ejection fraction with LVEF $\leq$ 40%, between 41 and 49%, and $\geq$50%, respectively.

Typical clinical presentation of heart failure includes impaired exercise capacity and fatigue. Participation in cardiac rehabilitation programs can improve physical fitness levels and alleviate symptoms. The beneficial outcomes of exercise interventions include [7] improved maximal oxygen uptake, an increase in peak work rate, and prolongation of the walking distance during 6MWT.

Exercise sessions commonly start during the inpatient phase once clinical stability is achieved. They involve progressive mobilization, strength training of small muscle groups, and respiratory exercises.

Phase II of cardiac rehabilitation typically starts at least a few weeks after a decompensation episode. It should be preceded by clinical examination, resting ECG, transthoracic echocardiography, and symptom-limited exercise testing [8]. Contraindications to exercise have been detailed in other sources. Important considerations and risks during exercise training include the following:

- -a weight increase over 2 kg within a few days during exercise,
- a systolic blood pressure reduction,
- ventricular arrhythmias,
- resting tachycardia,
- new episodes of atrial fibrillation,
- presence of ischemia during low-intensity exercise.

The exercise intensity should be ideally determined through cardiopulmonary testing (the gold standard for patients with heart failure), conducting low workload increments protocol (e.g., the Cornell, the Naughton for a treadmill or 5–10 W/min ramp protocol on a leg cycle ergometer). Based on physical performance, patients are classified as [9]:

- those with significant impairment of functional capacity (CPET: peak oxygen uptake ($VO_{2peak}$) below 10 mL $\times$ $kg^{-1}$ $\times$ $min^{-1}$, 6MWT distance (6MWD) < 300 m),
- those with moderate impairment of functional capacity (CPET: $VO_{2peak}$ 10–18 mL $\times$ $kg^{-1}$ $\times$ $min^{-1}$, 6MWD 300–450 m),
- those with good functional capacity (CPET: $VO_{2peak}$ > 18 mL $\times$ $kg^{-1}$ $\times$ $min^{-1}$, 6MWD > 450 m).

The initial duration and frequency should reflect the initial exercise capacity:

- < 3 METs: few sessions daily of 5–10 min,
- 3–5 MET 1–2 sessions daily of 15 min,
- > 5 MET: 1 session of 20–30 min.

Moderate-intensity continuous endurance training is recommended as a primary protocol with a duration of 20–40 min as a primary target, and exercise intensity increased subsequently. The training progression of an alternative interval approach is performed by changing the increased training duration up to 30 min, changing the work-to-recovery ratio from 1:2 to 1:1, or increasing active block intensity. Initial exercise protocol for poor physical capacity, i.e., low-intensity interval with 30-s active phases at 50% of workload and 1 min recovery phases, is recommended. After a few weeks, the transition from LIIT into a moderate-intensity continuous approach is advised.

Typical exercise training parameters are depicted below [10]:

- moderate intensity: 45–59% of VO2peak, 50–70% of peak work rate (PWR) (i.e., above the first ventilatory threshold), 40–59% of heart rate reserve (HRR), 4–6 METs, or 12–14 rate of perceived exertion (RPE) of the Borg scale. Further up-titration to moderate-to-high intensity for selected patients is advised (gradually reaching, if tolerated, the second ventilatory threshold (65–90% of peak oxygen uptake). The Talk Test could be considered a supportive tool when the heart rate is unreliable (e.g., atrial fibrillation), and the respiratory rate should allow for conversation.
- 2–3 sessions per week for outpatient and 4–5 sessions per week for residential program. Gradual increase following clinical status and symptoms to 3–5/week for outpatient and 5–6 for residential program is advised, with preferably all days per week finally.
- 15 to 30 min sessions, gradually increased to 40–60 min
- walking, Nordic walking, biking.

The high-intensity interval training may be applied to selected stable patients [11]. Typically, a 10-min warm-up phase is followed by short to moderate-time high-intensity intervals (above the second ventilatory threshold, RPE $\geq$ 15), interrupted by passive or active recovery intervals, attaining the final protocol of 4 $\times$ 4 min for selected patients.

Resistance training performed 2–3 times/week to increase muscle strength and aerobic capacity is strongly recommended [12]. An initial learning phase (<30% 1-RM, RPE 11–12, 5–10 repetitions) is followed by the maintenance phase with a higher number of repetitions and a low intensity (30–40% 1-RM, RPE 12), progressed up to 40–60% of the 1-RM (RPE > 15, 8–15 repetitions). Resistance training requires appropriate rest periods between sets, and a common work-to-rest ratio of 1:2, i.e., 30–60 s of exercise with a subsequent 1–2 min rest is recommended.

Resistance training progression (at least 4 weeks) should be phasic, as presented in Fig. 4.2.

Inspiratory muscle training is incorporated in those with inspiratory muscle weakness. Training starts at 30% of maximum inspiratory pressure and progresses up to a maximum of 60% by increasing the intensity every 7–10 days. Training commonly lasts 20–30 min/day with 3–5 weekly sessions for at least 2 months [13].

The neuromuscular electrical stimulation (NMES) may be used for patients with advanced heart failure. NMES improves exercise capacity, muscle strength, and quality of life [14].

The characteristics of exercise intervention for heart failure individuals are presented in Fig. 4.3.

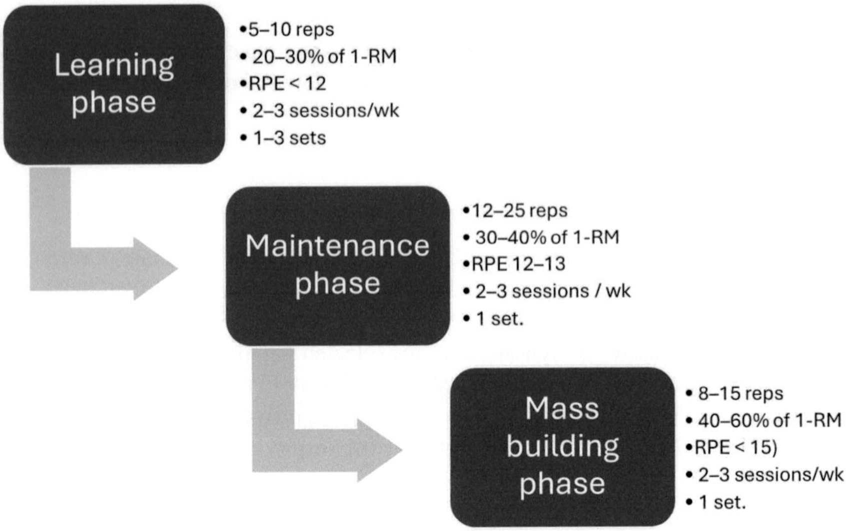


**Fig. 4.2** Progression of resistance training for heart failure individuals, by the authors RPE, rating of perceived exertion; 1-RM, one- repetition maximum

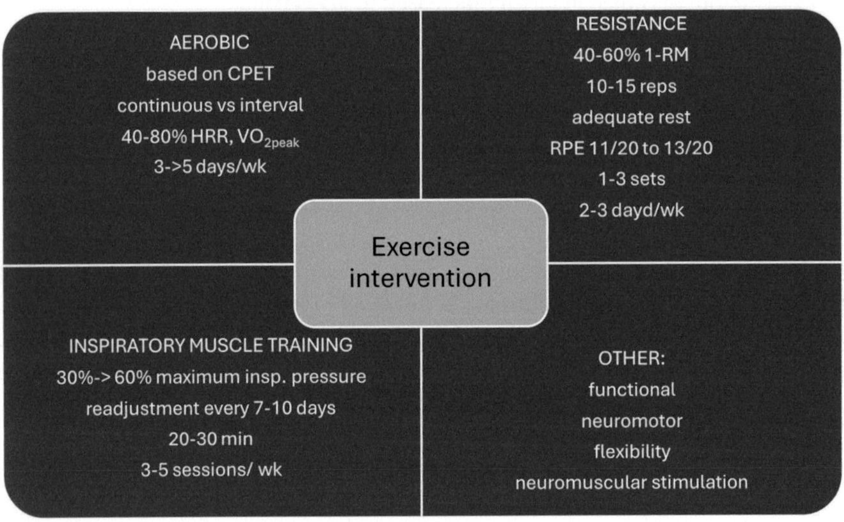


**Fig. 4.3** Exercise training program for patients with heart failure. By the authors CPET, cardiopulmonary exercise testing; HRR, heart rate reserve; RPE, rating of perceived exertion; VO2peak, peak oxygen uptake; wk, week; 1-RM, one- repetition maximum

## 4.4 Implantable Cardiac Electrical Devices

Many patients currently participating in cardiac rehabilitation programs have implanted cardiac devices, such as permanent pacemakers (PPM), cardiac resynchronization therapy devices (CRT), and implantable cardioverter-defibrillators (ICD). Exercise interventions for these patients are generally safe and effective; however, specific guidelines must be followed. The cardiac rehabilitation staff needs to have adequate knowledge about the functions of these devices. Patients in phase II of cardiac rehabilitation are typically aware that they should avoid extensive movements on the side of the implant for six weeks following the procedure to prevent lead fractures. As a result, upper-body strength exercises need to be modified, and resistance training is not allowed for 4 to 6 weeks after the device implantation.

Entry assessment should incorporate evaluation of underlying heart disease, potential device-related events if they occurred, and inspection of the wound. In the case of ICD or CRT with a defibrillator (CRT-D), to anticipate potential problems, knowledge about the detection threshold is mandatory (the upper limit of the training heart rate should be set between 10 and 20 bpm below the fire rate). Other device parameters to be interrogated include sensing, pacing threshold, and arrhythmia record [15].

A symptom-limited exercise test, preferably cardiopulmonary testing, is advised to provide information about chronotropic response to exercise, exertional arrhythmias, and maximum tracking rate. Exercise prescription for cardiac electrical device recipients with low left ventricular ejection fraction should follow the training rules for heart failure patients, with concern about the upper limits of training heart rate. It can include continuous or interval aerobic training following the progression described earlier. It is important to mention that seated arm exercises should be kept to a minimum, as it leads to reduced venous return, reduced end-diastolic volume, a concomitant decrease in cardiac output and an increased likelihood of arrhythmia. To reduce this impact, mild leg exercise, e.g., alternate heel raises, when combined with arm exercise, reduces the hemodynamic response compared with strict arm work.

Resistance training with 2–3 sets of 10–12 repetitions at 40–70% of one repetition maximum and a rate of perceived exertion of 12–15 is recommended [16]. Attention to shoulder movements on the side of the implant is obligatory.

The issues related to specific types of devices include [17]:

Permanent Cardiac Pacemaker:

- stimulation parameters
- symptoms related to incorrect stimulation, e.g., palpitations, dizziness, presyncope, syncope

Cardioverter-Defibrillator:

- history of ICD shocks
- relation of dysrhythmias and ICD shocks to exertion
- the upper limit of the training HR set 10–20 bpm below the fire rate

- exercise training is contraindicated in the presence of uncontrolled ventricular arrhythmias
- if ICD intervention occurs during exercise training, the session must be stopped and can be resumed after device control and clearance from the electrocardiologist. Exercise sessions should be resumed rapidly after a change in pharmacotherapy and/or the reprogramming of the device to avoid ICD discharge becoming a psychological barrier to physical activity.

Cardiac Resynchronization Therapy:

- training principles for heart failure; in the case of CRT-D, additional rules for ICD
- type of device inserted (CRT or CRT-D)
- the upper tracking limit value (stimulation with conduction 1:1)

The issues that must be considered when commencing an exercise training program for an individual with an implanted cardiac electrical device are presented in Fig. 4.4

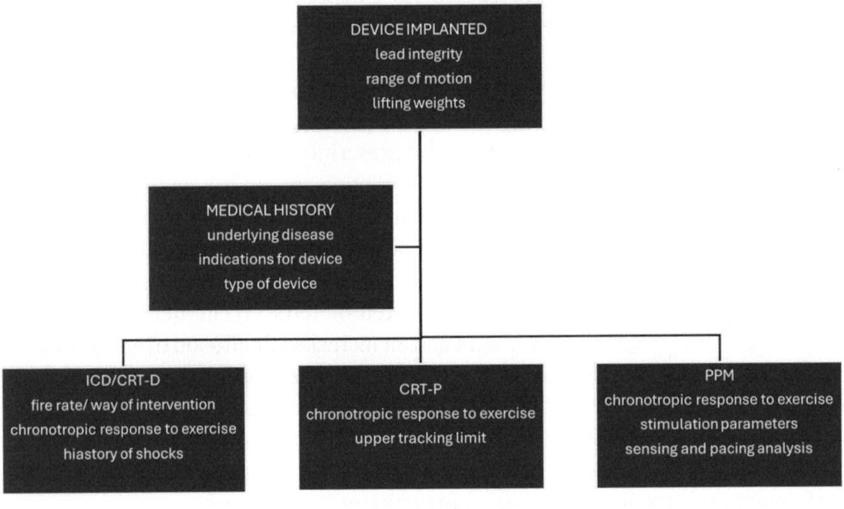


**Fig. 4.4** Cardiac implanted electrical device-related checklist, by the authors CRT, cardiac resynchronization therapy; CRT-D, cardiac resynchronization therapy with defibrillator; CRT-P, cardiac resynchronization therapy with pacing; ICD, implantable cardiac defibrillator; PPM, permanent cardiac pacemaker

## 4.5 Ventricular Assist Device

The left ventricular assist device (LVAD) serves as a bridge therapy to heart transplantation, recovery therapy, or destination therapy for those suffering from end-stage heart failure. The biventricular assist device (BiVAD) or LVAD allows patients to be discharged from the hospital setting.

Before discharge, the patient must be mobilized independently, performing essential activities of daily living and climbing stairs. The patient should be educated about the device, such as care, alarms, and power source changes.

Exercise can be safely prescribed if specific considerations follow it. Heart rate and blood pressure can be challenging to monitor due to the continuous flow pump, which eliminates the arterial pulse. Using a Doppler blood pressure monitor may be helpful in gaining more accurate readings, and the recommendation for the mean arterial pressure (MAP) should be 60 mmHg and above to start the exercise. Subjective measures are recommended to monitor exercise intensity, i.e., use of the RPE scale and symptoms such as fatigue and shortness of breath [18].

Exercise testing and exercise training are safe in patients with LVAD; however, patients can be severely deconditioned because of their previous status (before VAD). Cardiorespiratory fitness and hemodynamic response to exercise can be assessed by cardiopulmonary exercise testing, ECG exercise testing using an appropriate protocol for patients with low exercise capacity, and functional assessment with the 6-min walk test.

Exercise training improves the functional capacity of patients with VAD [19]. Due to the linear relationship between HR and VO2, an HRR of 60–80% as intensity can be used. In addition, maintaining a moderate level of perceived exertion (RPE 11–13) can be a good reflection of progression. Initially, patients should start with low intensity, as low as 2.5 METs, and monitor RPE and symptoms.

Due to wound healing, mild strengthening exercises can be implemented 8–12 weeks after device implantation. Other considerations include avoiding activities that increase intra-abdominal pressure and any contact sports that could damage the device. Patients should be appropriately informed about the device's alarms, power supply duration, and exercise intolerance signs and symptoms. Specific issues related to the VAD are presented in Fig. 4.5.

## 4.6 Heart Transplantation

Heart transplantation (HT) is associated with several physiological adaptations, including surgical denervation, decreased chronotropic competence, decreased ventricular function, pulmonary diffusion changes, decreased bone density, and changes in the morphology of the musculoskeletal system. Exercise training after heart transplantation improves aerobic capacity, muscle strength, and bone density.

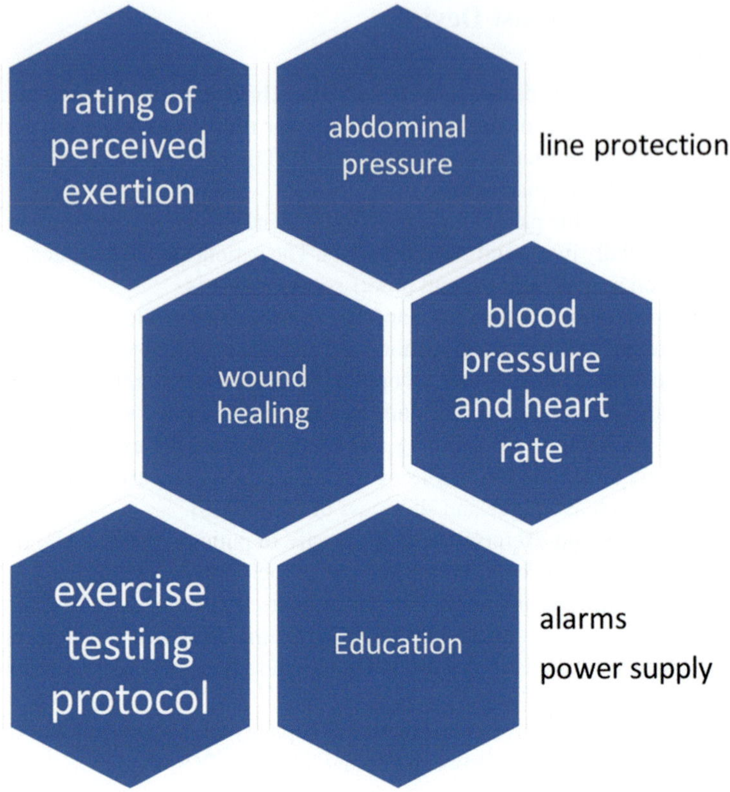


**Fig. 4.5** Issued related to exercise prescription for patients with a ventricular assist device

Both exercise testing and exercise training for the HT recipients require special attention. Because of heart denervation, resting heart rate (HR) is elevated, exertional HR increase is attenuated, and HR recovery is slow [20]. Thus, the highest HR may occur during exercise testing after the test or training session. Therefore, prolonged monitoring after the termination of testing or exercise is recommended.

Monitoring the rate of perceived exertion (RPE) is recommended (RPE). Aerobic training can be started in the second or third week after transplant, with resistance exercise added after 6–8 weeks. Prior to the commencement of the cardiac rehabilitation program, it is crucial to know the etiology of heart failure, details of immunotherapy, history of rejection episodes, and results of heart biopsies.

Cardiopulmonary testing is recommended as a gold standard for functional capacity assessment [21]. Testing protocol with small load increments is recommended, e.g., modified Bruce or Naughton. Aerobic training must include prolonged warm-up and cool-down phases considering the consequences of heart denervation (to enable catecholamines to increase and decrease the heart rate during both phases).

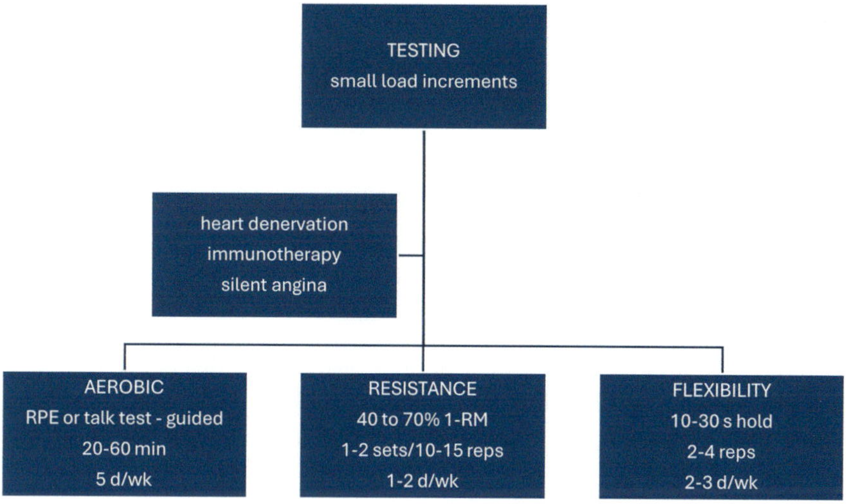


**Fig. 4.6** Exercise training for heart transplant recipients, by the authors 1-RM, one-repetition maximum; min, minute; reps, repetitions; RPE, rating of perceived exertion; wk, week

As a result of the denervation, patients do not feel the symptoms of ischemia. Therefore, a close ECG monitoring during training is important. As mentioned earlier, the rating of perceived exertion and respiratory rate should guide exercise intensity rather than specific heart rate. In the case of steroid therapy for rejection episodes, steroid-induced myopathy can occur. In this case, exercise intensity may need to be kept lower (RPE 11–13) initially to allow for muscle adaptations.

The characteristics of the exercise training program after heart transplantation are depicted in Fig. 4.6.

## 4.7 Peripheral Arterial Disease

Peripheral arterial disease (PAD) limits blood flow and oxygen supply to the lower limbs, resulting in dysfunction of the skeletal muscles. The hallmark of PAD is intermittent claudication, manifesting in 30% of patients [22]. The spectrum of positive changes made by regular walking programs includes symptom reduction, increased claudication distance, and quality of life improvement. However, there is no evidence of the influence on mortality reduction. Postulated mechanisms of exercise tolerance improvement include enhanced vasodilation, blood redistribution, reduced viscosity, better walking patterns, and increased muscle strength.

Exercise testing determines free maximal pain-free distance; testing protocol typically begins with low speed and gradual increments in treadmill grade to induce the onset of claudication within 3–5 min and moderate to severe claudication level within 10 min. Time and distance to the onset of claudication and cessation of walking due to unbearable pain should be recorded [23]. Exercise test also allows for the detection of coronary ischemia or chronotropic incompetence.

The most popular testing protocols include [24]:

- Hiatt protocol (constant speed of 3.2 km/h, with 3.5% grade change every 3 min
- Gardner-Skinner protocol (constant speed of 3.2 km/h and grade change by 2% every 2 min).

The ankle-brachial index (ABI), i.e., the ratio of the blood pressures at the ankle and upper arm level, is a useful tool for determining PAD advancement level. ABI following exercise in PAD individuals is reduced because blood flow is switched into the proximal area of the leg. The ABI below 0.5 reflects significant stenosis. Exertional ABI decrease precedes claudication, and another sign of impaired lower limb blood supply is delayed return of blood pressure value to the pre-exertional [25].

Exercise training intensity can be determined based on the following:

- claudication
- maximal walking distance

Intermittent bouts of walking are the most common exercise training forms for patients with PAD [26]. Training intensity is set to match the selected level of discomfort of the lower limbs, with exercise/rest bouts repeated throughout the session. The training session starts with 5–10 min of warm-up and ends with 5–10 min of cool-down phase. Most of the exercise training protocols incorporate walking till moderate pain or even till near maximal pain, with a subsequent rest period.

The commonly used training protocols entail:

- speed of 4.6 km/h and grade adjustment not to provoke pain within 3 min speed of 4.0 km/h without inclination and with speed manipulation only to walk 10 min without claudication

Although weight-bearing exercise is the recommended mode of exercise for patients with PAD, leg cycle ergometers may be valuable alternatives for obese individuals and those with knee problems [27]. For a patient without intermittent claudication, an exercise training intensity of 40–50% of heart rate reserve (HRR) is recommended, and it should be progressed to 70–80% HRR within 6 months.

Resistance training commonly includes three sets of eight repetitions of exercises for seven different muscle groups. The training intensity starts at 20–50% of 1-RM, and it is recommended to progress up to 80% of 1-RM over a few weeks.

The overall cardiac rehabilitation program duration should last at least 12 weeks and, optimally, six months. The rest periods between walking intervals should allow for full pain relief.

The apparent obstacle for this population remains long-term adherence. Moreover, provoking pain raises questions about inflammatory endothelial damage. Maximal vasodilation in lower limb muscles occurs before claudication. Therefore, alternative approaches in the form of pain-free walking protocols have been postulated [28]. Such a method can enhance long-term adherence and protect against local inflammation.

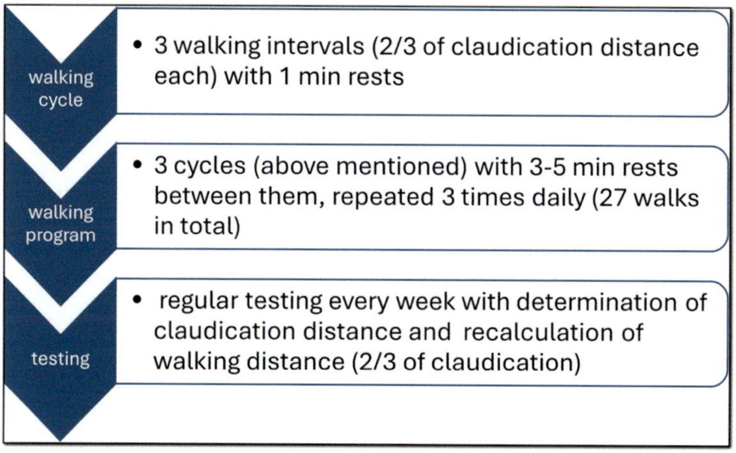


**Fig. 4.7** The pain-free walking protocol principle by the authors

One such alternative approach has been utilized by the authors, known as the "2/3 of claudication formula". It consists of 3–5 bouts of walking (covering 2/3 of claudication distance) separated by 1 min rest. Such a cycle is repeated 3 times with 3–5 min breaks between the cycles ($3 \times 3$ walks), and such training should be performed three times daily.

The concept of this is presented in Fig. 4.7. The progression is based on weekly re-testing with the determination of claudication distance.

The FITT formula of exercise training and training pathways has been presented in Fig. 4.8.

General approach to exercise training prescription for patients with PAD is depicted in Fig. 4.9

## 4.8 Frailty, Elderly

Exercise training for older people should maintain mobility and independence and prevent frailty. Older people present a spectrum of abnormalities: they are prone to falls, with a high prevalence of osteoarthritis, chronic obstructive pulmonary disease, dementia, impaired sight, hearing, balance, and urine incompetence [29]. Therefore, the exercise program requires detailed preassessment by a multidimensional geriatric assessment (MGA). MGA rules out disability, cognitive problems, or frailty, as these conditions require a specific approach [30].

The MGA measures incorporate the following:

- basic activities of daily living, e.g., Katz activities of daily living for assessment of self-care and mobility-instrumental activities of daily living, i.e., activities such as shopping or cooking

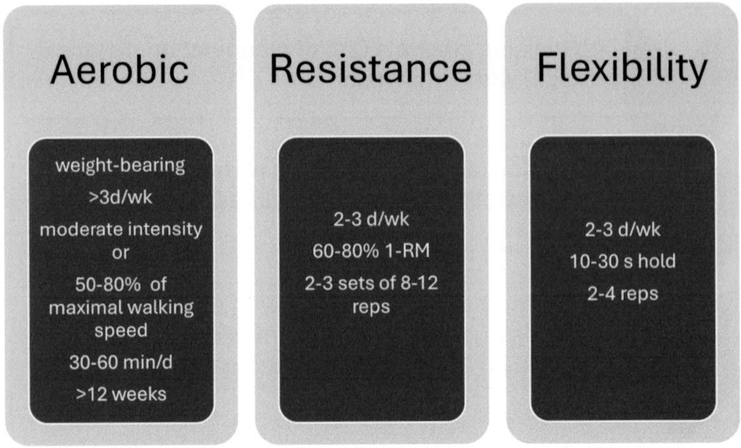


**Fig. 4.8** FITT principle for peripheral arterial disease patients, by the authors FITT, frequency, intensity time and type; 1-RM, one-repetition maximum; d, day; min, minute; reps, repetitions; wk, week

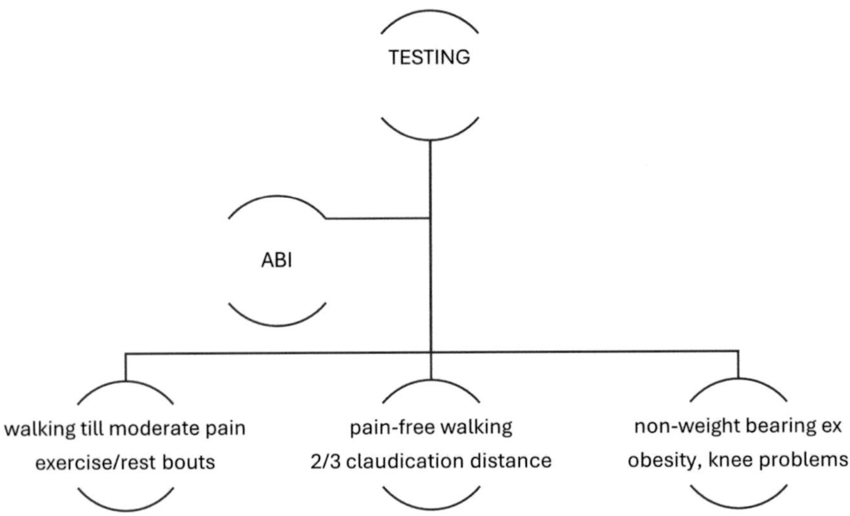


**Fig. 4.9** Peripheral arterial disease aerobic training pathway, by the authors ABI, ankle-brachial index; PAD, peripheral arterial disease

- social support
- mental status, i.e., presence of anxiety or depression; cognitive function
- balance and gait
- nutritional status
- grip strength
- comorbidities

The most used functional assessment tool is the Senior Fitness Test (Fullerton Test). Other commonly performed physical performance tests include the Short Physical Performance Battery (SPPB), usual gait speed, and continuous-scale physical performance test [31].

The Senior Fitness Test includes seven tasks: 30-s chair stand, 3 s arm curl, 8 ft up and go, 6-min walk, 2-min step test, sit and reach, and back scratch.

The 30-s chair stand and the arm curl are designed to evaluate lower and upper body strength, respectively. The chair sit-and-reach and back scratch tests target flexibility of the lower and upper body. The eight-foot and go test evaluates balance, and the six-minute walk test and the 2-min step test evaluate endurance.

The frailty is described as impairment in physical, psychological, and social abilities. As frailty is considered a strong predictor of mortality, morbidity, and hospitalization, several screening models for frailty have been developed, with the above-mentioned MGA, the phenotypic Freid model, and the deficit accumulation model widely adapted [32]. The Freid model targets nutrition, subjective assessment of physical exhaustion, energy expenditure, mobility, and muscle strength. The frailty is recognized with three or more abnormalities present.

Exercise testing should start with a load of <3 MET and small (i.e., 0.5–1.0 MET) workload increments (e.g., the modified Naughton protocol with an increase in grade rather than in speed). Leg cycle ergometers should be considered for testing those with balance or gait problems. Many frail patients are unable to perform a standard exercise stress test. Therefore, the training intensity applied should be set at a heart rate slightly lower than that attained during 6MWT [33].

Exercise prescription should target endurance, strength, flexibility, coordination, and balance exercises. Endurance training should be started at low intensity, with the first sessions lasting 10–15 min. The warm-up and cool-down phases can be prolonged. Moderate-intensity training can be beneficial at later stages. The Borg scale and talk test rather than heart rate should guide the workload for patients with atrial fibrillation. Program duration is often extended beyond 12 weeks [34].

Resistance training on non-consecutive days at light-to-moderate intensity (30–70% of 1-RM) progressed, if tolerated, to moderate–high intensity (70–85% 1-RM), with 8–12 repetitions involving 6–8 groups of skeletal muscles. Multi-weight machines are useful for stabilizing body position while lifting weights (36).

Neuromotor training merging balance, agility, and proprioceptive training should be performed 2–3 times weekly and include complex postures, e.g., tai chi [35]. In severely frail patients, the volume of exercise should be reduced. The exercise regimen is often limited to bed mobilization only.

The exercise training principle has been presented in Figs. 4.10 and 4.11.

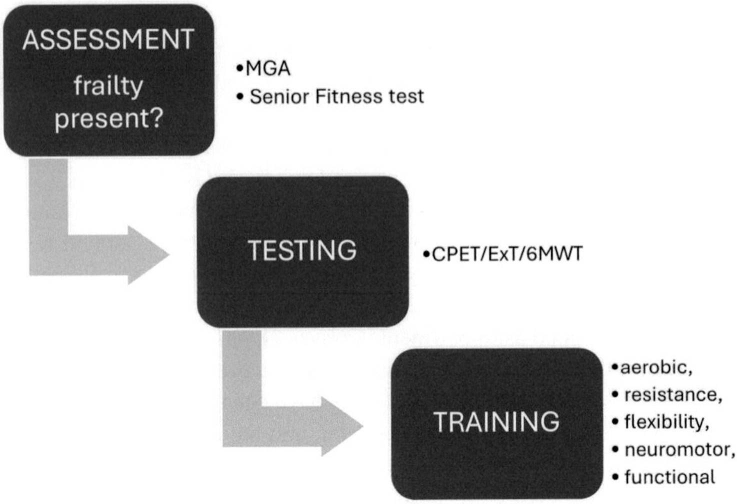


**Fig. 4.10** Exercise intervention pathway for the elderly, by the authors MGA, multidimensional geriatric assessment; CPET, cardio-pulmonary exercise test; ExT, exercise test; 6MWT, six-minute walk test

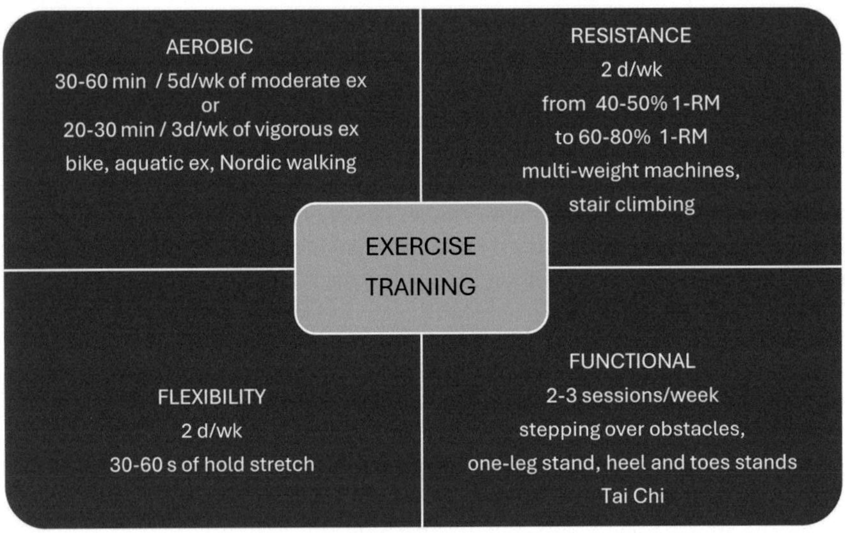


**Fig. 4.11** Components of exercise training program for the elderly, by the authors ex, exercises; 1-RM, one-repetition maximum; wk, week

## 4.9 Patients with Comorbidities

### *4.9.1 Diabetes Mellitus*

Diabetes mellitus (DM) is an effect of insufficient insulin secretion or inappropriate insulin usage. Physical activity prevents complications related to diabetes and increases insulin sensitivity. The principle of exercise training is like these applicable to general population, however, diabetes complications may require modifications to the training program, e.g., adjustment of resistance training in case of retinopathy. Considerations for diabetes-related issues when designing exercise training programs are outlined in Fig. 4.12.

In general, it is recommended to exercise at least 3–5 days per week for at least 30 min per session at a moderate-to-high intensity. The evidence is growing that high-intensity interval training may be more effective in lowering blood HbA1c than a moderate-intensity aerobic program. The resistance training has a positive impact on reducing HbA1c level.

Exercise training should not provoke hypoglycaemia during or after exercise. Thus, blood glucose monitoring is crucial. Recently introduced antidiabetic medications are better tolerated and should be preferred. Strict blood glucose level monitoring is recommended for patients taking insulin to avoid hypoglycaemia; therefore,

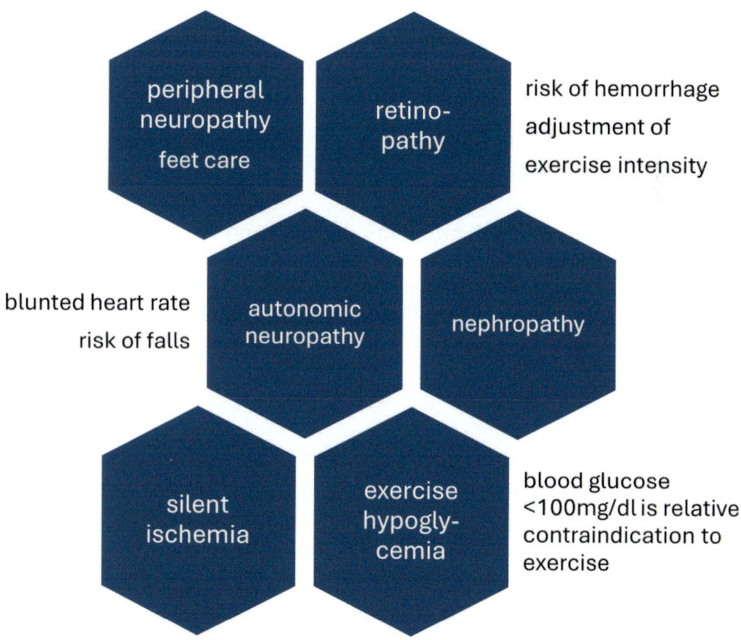


**Fig. 4.12** Diabetes mellitus-related issues affecting exercise training program, by the authors

before starting the exercise program, the patient should be checked for a recent history of glycaemic control. In the case of therapy on insulin, adjusting the insulin dose and its administration time depending on the planned exercise is crucial. In case of glucose level <100 mg/dl before exercise session, the supply of carbohydrates with subsequent glycaemic control is recommended. Consumption of a carbohydrate snack minimizes the risk of hypoglycaemia. Training should be stopped immediately in case of hypoglycaemic episodes during exercise sessions, and patient should consume easily digestible carbohydrates.

### 4.9.2 Obesity

Obesity is a known risk factor for developing cardiovascular diseases and results in musculoskeletal problems [36]. Effective obesity treatment depends on energy balance, i.e., energy expenditure must exceed energy intake. Research suggest that at least 3–5% of weight loss is required to significantly reduce cardiovascular risk factors [37] The exercise intervention itself is not enough for weight loss, thus, combined approach, that includes diet, and exercise is the most effective strategy. To maximize benefits, weekly aerobic exercise volume should ideally reach 250 min, and energy expenditure of at least 1500 kcal should be achieved [38].

Moderate-intensity exercise, e.g., walking, allows for achieving exercise training volume, with more frequent sessions as another option. Individuals with obesity should increase their habitual activity level, and online applications that allow for counting steps, kilocalories, and distance covered could be helpful. Low-intensity aerobic training is modestly affecting energy expenditure and loss of body mass.

Strength training at a maximum of 60–70% of one repetition, 3 sets of 12–15 repetitions each, will increase muscle mass and strength and improve the lipid profile [39].

The specific obesity-related problems during exercise training programs are presented in Fig. 4.13.

### 4.9.3 Arterial Hypertension

Arterial hypertension leads to cardiovascular diseases and premature death. Genetic predisposition and lifestyle account for primary hypertension, representing 95% of all cases, whereas secondary hypertension is caused by specific medical conditions, with chronic kidney disease, renal artery stenosis and pheochromocytoma contributing mostly [40].

Research shows that regular physical activity can lower blood pressure by 5–7 mmHg in individuals with hypertension during submaximal exertion. There is a dose–response relationship between the volume of exercise and blood pressure levels,

**Fig. 4.13** Obesity-related issues affecting exercise intervention outcomes, by the authors

provided that the exercise is of at least moderate intensity and the duration of each session is between 30 to 45 min.

Resistance training reduces blood pressure to at least the same magnitude as aerobic exercises. To avoid excessive blood pressure increases, resistance training should be started at low intensity [41]. During strengthening exercises, the patients should be encouraged to breathe and avoid the Valsalva manoeuvre. Figure 4.14 depicts general exercise training rules for individuals with hypertension.

### *4.9.4 Dyslipidaemia*

Dyslipidaemia is characterized by elevated levels of total cholesterol, low-density lipoprotein (LDL) and triglycerides (TG), and reduced levels of high-density lipoproteins (HDL). An aerobic exercise modifies the lipid profile with a higher impact on TG and HDL and modest LDL reduction (by 3–6 mg/dl) [42]. Resistance training has an additional positive impact on lipid profile, but the studies' results are inconsistent. Statins remain the most potent treatment for dyslipidaemia, with dosage adjusted to reduce LDL below a specific threshold depending on the patient's risk

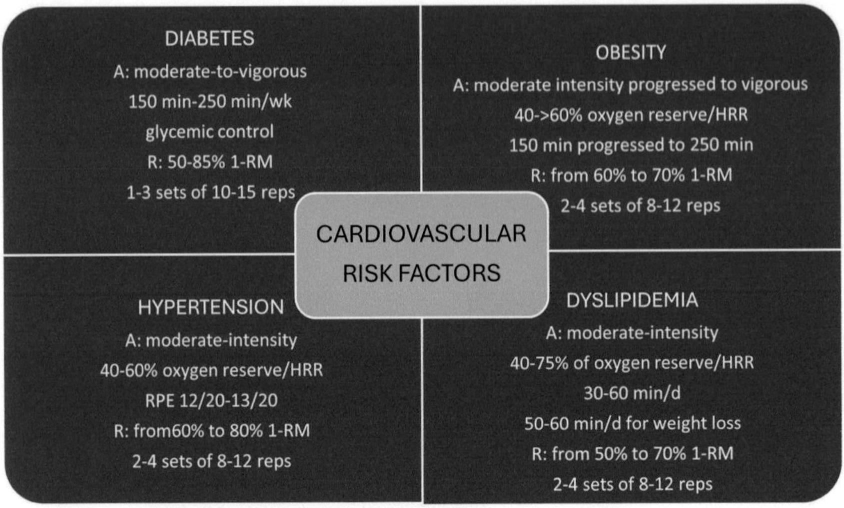


**Fig. 4.14** The exercise training principle for individuals with cardiovascular risk factors, by the authors. A, aerobic training; R, resistance training; 1-RM, one-repetition maximum; HRR, heart rate reserve; RPE, rate of perceived exertion; reps, repetitions

profile. The combination of pharmacotherapy with dietary modification and exercise training intervention is recommended as the most effective therapy. In case of statin-induced myopathy, statin dose should be modified accordingly, and ezetimibe and PCSK9-inhibitors should be considered. General exercise training rules are like those for healthy individuals; however, to achieve a positive modification in lipid profile, energy expenditure of at least 900 kcal per week is required [43].

The exercise training principle for individuals with dyslipidaemia is summarized in Fig. 4.14.

**AuthorContribution Competing Interests** The authors have no conflicts of interest to declare that are relevant to the content of this chapter.

## References

1. Lee JY, Han S, Ahn JM, Park DW, Kang SJ, Lee SW, et al. Impact of participation in phase I and phase II cardiac rehabilitation on long-term survival after coronary artery bypass graft surgery. Int J Cardiol. 2014;176:1429–32. https://doi.org/10.1016/j.ijcard.2014.08.041.
2. Sibilitz KL, Berg SK, Tang LH, Risom SS, Gluud C, Lindschou J, et al.: Exercise-based cardiac rehabilitation for adults after heart valve surgery. Cochrane Database Syst Rev. 2016;3:CD010876. https://doi.org/10.1002/14651858.
3. Pressler A, Christle JW, Lechner B, Grabs V, Haller B, Hettich I, et al. Exercise training improves exercise capacity and quality of life after transcatheter aortic valve implantation: a randomized pilot trial. Am Heart J. 2016;182:44–53. https://doi.org/10.1016/j.ahj.2016.08.00.
4. Abreu A, Frederix I, Dendale P, Janssen A, Doherty P, Piepoli MF, et al. Secondary prevention and rehabilitation section of EAPC reviewers: Standardization and quality improvement of

secondary prevention through cardiovascular rehabilitation programmes in Europe: the avenue towards EAPC accreditation programme: a position statement of the secondary prevention and rehabilitation section of the european association of preventive cardiology (EAPC). Eur J Prev Cardiol. 2021;28:496–509. https://doi.org/10.1177/2047487320924912.
5. La Gerche A. The potential cardiotoxic effects of exercise. Can J Cardiol. 2016;32:421–8. https://doi.org/10.1016/j.cjca.2015.11.010.
6. McDonagh TA, Metra M, Adamo M, Gardner RS, Baumbach A, Böhm M, et al.: ESC scientific document group. 2023 focused update of the 2021 ESC guidelines for the diagnosis and treatment of acute and chronic heart failure. Eur Heart J. 2023;44:3627–3639. https://doi.org/10.1093/eurheartj/ehad195.
7. Belardinelli R, Georgiou D, Cianci G, Purcaro A. Randomized, controlled trial of long-term moderate exercise training in chronic heart failure: effects on functional capacity, quality of life, and clinical outcome. Circulation. 1999;99:1173–82. https://doi.org/10.1161/01.cir.99.9.1173.
8. Cunha-Rocha F, Rocha JA. When and how to start exercise training after heart failure decompensation? Still more questions than answers. Eur J Prev Cardiol. 2018;25:1899–900. https://doi.org/10.1177/2047487318797393.
9. Piepoli MF, Conraads V, Corrà U, Dickstein K, Francis DP, Jaarsma T, et al.: Exercise training in heart failure: from theory to practice. A consensus document of the heart failure association and the european association for cardiovascular prevention and rehabilitation. Eur J Heart Fail. 2011;13:347–57. https://doi.org/10.1093/eurjhf/hfr017.
10. Corrà U, Giannuzzi P, Adamopoulos S, Bjornstad H, Bjarnason-Weherns B, Cohen-Solal A, et al.: Working group on cardiac rehabilitation and exercise physiology of the European society of cardiology. Executive summary of the position paper of the working group on cardiac rehabilitation and exercise physiology of the European society of cardiology (ESC): core components of cardiac rehabilitation in chronic heart failure. Eur J Cardiovasc Prev Rehabil. 2005;12:321–5. https://doi.org/10.1097/01.hjr.0000173108.76109.88.
11. Meyer K, Samek L, Schwaibold M, Westbrook S, Hajric R, Lehmann M, Essfeld D, Roskamm H. Physical responses to different modes of interval exercise in patients with chronic heart failure–application to exercise training. Eur Heart J. 1996;17:1040–7. https://doi.org/10.1093/oxfordjournals.eurheartj.a015000.
12. Williams MA, Haskell WL, Ades PA, Amsterdam EA, Bittner V, Franklin BA, et al.: American heart association council on clinical cardiology; American heart association council on nutrition, physical activity, and metabolism. Resistance exercise in individuals with and without cardiovascular disease: 2007 update: a scientific statement from the American Heart association council on clinical cardiology and council on nutrition, physical activity, and metabolism. Circulation. 2007;116:572–84.https://doi.org/10.1161/CIRCULATIONAHA.107.185214.
13. Laoutaris I, Dritsas A, Brown MD, Manginas A, Alivizatos PA, Cokkinos DV. Inspiratory muscle training using an incremental endurance test alleviates dyspnea and improves functional status in patients with chronic heart failure. Eur J Cardiovasc Prev Rehabil. 2004;11(6):489–96. https://doi.org/10.1097/01.hjr.0000152242.51327.63.
14. Saitoh M, Dos Santos MR, Anker M, Anker SD, von Haehling S, Springer J. Neuromuscular electrical stimulation for muscle wasting in heart failure patients. Int J Cardiol. 2016;225:200–5. https://doi.org/10.1016/j.ijcard.2016.09.127.
15. Haennel RG. Exercise rehabilitation for chronic heart failure patients with cardiac device implants. Cardiopulm Phys Ther J. 2012;23:23–8.
16. Iliou MC, Blanchard JC, Lamar-Tanguy A, Cristofini P, Ledru F. Cardiac rehabilitation in patients with pacemakers and implantable cardioverter defibrillators. Monaldi Arch Chest Dis. 2016;86:756. https://doi.org/10.4081/monaldi.2016.756.
17. Pedretti RFE, Iliou MC, Israel CW, Abreu A, Miljoen H, Corrà U, et al. Comprehensive multicomponent cardiac rehabilitation in cardiac implantable electronic devices recipients: a consensus document from the European Association of Preventive Cardiology (EAPC; Secondary prevention and rehabilitation section) and European Heart Rhythm Association (EHRA). Eur J Prev Cardiol. 2021;28:1736–52. https://doi.org/10.1093/eurjpc/zwaa121.

18. Probert H, BarriTt H, Breen S, Buckley J, Burgess L, Graham K, Green F, Grove T, Turner S, Warren C. ACPICR STANDARDS: standards of physical activity and exercise in the cardiovascular population. 2015. 3rd edition.
19. Mahfood Haddad T, Saurav A, Smer A, Azzouz MS, Akinapelli A, Williams MA, Alla VM. Cardiac rehabilitation in patients with left ventricular assist device: a systematic review and meta-analysis. J Cardiopulm Rehabil Prev. 2017;37:390–6. https://doi.org/10.1097/HCR.000 0000000000254.
20. Kavanagh T, Yacoub MH, Mertens DJ, Kennedy J, Campbell RB, Sawyer P. Cardiorespiratory responses to exercise training after orthotopic cardiac transplantation. Circulation. 1988;77(1):162–71. https://doi.org/10.1161/01.cir.77.1.162.
21. Anderson L, Nguyen TT, Dall CH, Burgess L, Bridges C, Taylor RS. Exercise-based cardiac rehabilitation in heart transplant recipients. Cochrane Database Syst Rev. 2017;4:CD012264. https://doi.org/10.1002/14651858
22. McDermott MM, Hoff F, Ferrucci L, Pearce WH, Guralnik JM, Tian L, et al. Lower extremity ischemia, calf skeletal muscle characteristics, and functional impairment in peripheral arterial disease. J Am Geriatr Soc. 2007;55:400–6. https://doi.org/10.1111/j.1532-5415.2007.01092.x.
23. Tan KH, Cotterrell D, Sykes K, Sissons GR, de Cossart L, Edwards PR. Exercise training for claudicants: changes in blood flow, cardiorespiratory status, metabolic functions, blood rheology and lipid profile. Eur J Vasc Endovasc Surg. 2000;20:72–8. https://doi.org/10.1053/ejvs.2000.1137.
24. Riebe D, Patterson RB, Braun CM. Comparison of two progressive treadmill tests in patients with peripheral arterial disease. Vasc Med. 2001;6:215–21. https://doi.org/10.1177/1358836x0 100600403.
25. Alqahtani KM, Bhangoo M, Vaida F, Denenberg JO, Allison MA, Criqui MH. Predictors of change in the ankle brachial index with exercise. Eur J Vasc Endovasc Surg. 2018;55:399–404. https://doi.org/10.1016/j.ejvs.2017.12.004.
26. Lane R, Ellis B, Watson L, Leng GC. Exercise for intermittent claudication. Cochrane Database Syst Rev. 2014;(7):CD000990. https://doi.org/10.1002/14651858.CD000990.pub3.
27. Gardner AW, Katzel LI, Sorkin JD, Goldberg AP. Effects of long-term exercise rehabilitation on claudication distances in patients with peripheral arterial disease: a randomized controlled trial. J Cardiopulm Rehabil. 2002;22:192–8. https://doi.org/10.1097/00008483-200205000-00011.
28. Mika P, Spodaryk K, Cencora A, Unnithan VB, Mika A. Experimental model of pain-free treadmill training in patients with claudication. Am J Phys Med Rehabil. 2005;84:756–62. https://doi.org/10.1097/01.phm.
29. Rich MW, Chyun DA, Skolnick AH, Alexander KP, Forman DE, Kitzman DW, et al. Knowledge gaps in cardiovascular care of older adults: a scientific statement from the American heart association, American college of cardiology, and American geriatrics society: executive summary. J Am Geriatr Soc. 2016;64:2185–92. https://doi.org/10.1111/jgs.14576.
30. Stortecky S, Schoenenberger AW, Moser A, Kalesan B, Jüni P, Carrel T, et al. Evaluation of multidimensional geriatric assessment as a predictor of mortality and cardiovascular events after transcatheter aortic valve implantation. JACC Cardiovasc Interv. 2012;5:489–96. https://doi.org/10.1016/j.jcin.2012.02.012.
31. Rikli RE, Jones CJ. Development and validation of criterion-referenced clinically relevant fitness standards for maintaining physical independence in later years. Gerontologist. 2013;53:255–67. https://doi.org/10.1093/geront/gns071.
32. Xue QL. The frailty syndrome: definition and natural history. Clin Geriatr Med. 2011;27:1–15. https://doi.org/10.1016/j.cger.2010.08.009.
33. Afilalo J, Alexander KP, Mack MJ, Maurer MS, Green P, Allen LA, et al. Frailty assessment in the cardiovascular care of older adults. J Am Coll Cardiol. 2014;63(8):747–62. https://doi.org/10.1016/j.jacc.2013.09.070.
34. Theou O, Stathokostas L, Roland KP, Jakobi JM, Patterson C, Vandervoort AA, Jones GR. The effectiveness of exercise interventions for the management of frailty: a systematic review. J Aging Res. 2011;2011: 569194. https://doi.org/10.4061/2011/569194.

35. Taylor-Piliae RE. Tai Chi as an adjunct to cardiac rehabilitation exercise training. J Cardiopulm Rehabil. 2003;23:90–6. https://doi.org/10.1097/00008483-200303000-00004.
36. Biener A, Cawley J, Meyerhoefer C. The impact of obesity on medical care costs and labor market outcomes in the US. Clin Chem. 2018;64:108–17. https://doi.org/10.1373/clinchem.2017.272450.
37. Jensen MD, Ryan DH, Apovian CM, Ard JD, Comuzzie AG, Donato KA, et al.: American college of cardiology/American heart association task force on practice guidelines; obesity society. 2013 AHA/ACC/TOS guideline for the management of overweight and obesity in adults: a report of the American college of cardiology/American heart association task force on practice guidelines and the obesity society. Circulation. 2014;129(25 Suppl 2):S102–38. https://doi.org/10.1161/01.cir.0000437739.71477.ee.
38. Donnelly JE, Blair SN, Jakicic JM, Manore MM, Rankin JW, Smith BK; American college of sports medicine. American college of sports medicine position stand. Appropriate physical activity intervention strategies for weight loss and prevention of weight regain for adults. Med Sci Sports Exerc. 2009;41:459–71. https://doi.org/10.1249/MSS.0b013e3181949333.
39. Curioni CC, Lourenço PM. Long-term weight loss after diet and exercise: a systematic review. Int J Obes (Lond). 2005;29:1168–74. https://doi.org/10.1038/sj.ijo.0803015.
40. Rosendorff C, Lackland DT, Allison M, Aronow WS, Black HR, Blumenthal RS, et al.: American heart association, American college of cardiology, and American society of hypertension. Treatment of hypertension in patients with coronary artery disease: a scientific statement from the American Heart Association, American college of cardiology, and American society of hypertension. Hypertension. 2015;65:1372–407. https://doi.org/10.1161/HYP.0000000000000018.
41. Pescatello LS, Buchner DM, Jakicic JM, Powell KE, Kraus WE, Bloodgood B, et al.: 2018 Physical activity guidelines advisory committee*. Physical activity to prevent and treat hypertension: a systematic review. Med Sci Sports Exerc. 2019;51:1314–1323. https://doi.org/10.1249/MSS.0000000000001943.
42. Wang Y, Xu D. Effects of aerobic exercise on lipids and lipoproteins. Lipids Health Dis. 2017;16:132. https://doi.org/10.1186/s12944-017-0515-5.
43. Hansen D, Niebauer J, Cornelissen V, Barna O, Neunhäuserer D, Stettler C, et al. Exercise prescription in patients with different combinations of cardiovascular disease risk factors: a consensus statement from the expert working group. Sports Med. 2018;48:1781–97. https://doi.org/10.1007/s40279-018-0930-4.

# Chapter 5
# Future Directions

**Adam Staron and Jerzy Rybicki**

**Abstract** Individualized physical activity prescription should be strongly recommended after completing a cardiac rehabilitation program as well as for individuals with cardiovascular risk factors. Such a long-term intervention with a goal identified and regular reassessment will ensure a patient's adherence and amplify his/her motivation. The utilization of the existing guidelines on physical activity for individuals with increased cardiovascular risk is still unsatisfactory. Thus, efforts are being made to implement feasible tools for exercise prescription, such as the EXPERT tool, customized exercise intensity calculators, or individually tailored exercise prescriptions on discharge from the cardiac center. With the growing number of exercise activity trackers and programming tools, the next step, in our opinion, should be establishing exercise centers, binding primary and secondary cardiovascular prevention, and supported by meticulous functional capacity and myocardial function assessment. The only way to rejuvenate and well-being is to understand the role of regular physical activity and dietary and psychosocial support for the cardiovascular system.

**Keywords** Artificial intelligence · Holistic approach · Exercise prescription · Exercise center · Cardiovascular prevention

## 5.1 Exercise Prescription on Discharge

The recommendations for physical activity suggest engaging in exercise training at least three to five times per week, ideally every day. The training volume, similar to the medication dosage, is crucial for effective exercise prescription. It is advised to accumulate at least 30 min of moderate-intensity exercise per day (a total of at least

A. Staron (✉)
The Upper Silesian Medical Center Katowice Poland, Katowice, Poland
e-mail: astaron1973@wp.pl

J. Rybicki
"REPTY" Upper Silesian Rehabilitation Center Tarnowskie Góry Poland, Silesian Center for Heart Disease, Zabrze, Poland
e-mail: j_rybicki@cyberia.pl

 
J. Rybicki and A. Staron (eds.), *Exercise Programming for Cardiac Patients*,
https://doi.org/10.1007/978-3-031-88475-7_5

150 min per week), or at least 75 min of vigorous-intensity exercise per week, or a combination of both [1]. However, low adherence to these physical activity guidelines indicates a need for strategies to overcome potential barriers. Incorporating feasible and motivating solutions, such as interactive digital tools is important. A specific strategy for promoting long-term physical activity is to provide exercise prescriptions upon discharge from the acute cardiac phase or a cardiac rehabilitation center. Given the low referral rates, poor adherence to cardiac rehabilitation programs, and barriers faced by certain groups (including women, older adults, and minorities), personalized exercise prescriptions provided at discharge can be a viable solution. Such a prescription, tailored to an individual's needs, should be progressive, safe, and include follow-up and monitoring. Prescribed exercise programs must specify the type, intensity, duration, and frequency of exercise sessions. The specific cardiac condition, clinical status, and ongoing medical therapy must also be considered [2]. The intensity of exercise sessions should be set between the first and second ventilatory thresholds. For those with known ischemia or arrhythmia, exercise intensity should be kept below the event thresholds. Calculating the training volume in the time unit (e.g., weekly) and setting it as a goal can be beneficial. An example of an aerobic training proposal (using an interval-to-continuous protocol) is provided in Table 5.1.

Incorporation of a long-term strategy requires adequate safety tools. Thus, utilization of currently available devices capable of monitoring heart rate, energy expenditure, steps, and distance covered is crucial. A more sophisticated strategy has been offered by the European Association of Preventive Cardiology.

The EXPERT tool is an interactive digital tool designed for exercise specialists [3]. The EXPERT system incorporates exercise prescription algorithm utilizing current European guidelines and position statements. EXPERT tool delivers recommendations based on analysis of the physical capacity (ideally through cardiopulmonary exercise test), current pharmacotherapy and history of adverse events during exercise testing. The EXPERT tool serves both as recommendation tool and training application [4]. The former assists professionals during exercise training, whereas the latter offers clinical scenarios for effective exercise prescription.

Customized exercise intensity calculators have emerged as another promising tool for long-term exercise prescription.

Another useful tool is the training heart rate (or load) calculator, the MET-to-Watt converter, which translates peak MET on the treadmill into estimated peak work rate on the bike (Fig. 5.1), or the treadmill incline-grade converter, which prescribes the optimal speed and inclination for treadmill walking training (Fig. 5.2).

The advanced digital solutions are customized to support exercise prescriptions based on data derived from cardiopulmonary exercise tests or ECG exercise tolerance tests. This testing data is transformed into a personalized exercise intensity calculator, which provides optimal training parameters and suggested exercise protocols. Additionally, the system can act as an exercise tutorial, allowing for instant calculations of specific training parameters. An example of this personalized approach, which is founded on exercise testing and offers a variety of individually tailored exercise training protocols, is illustrated in Fig. 5.3.

**Table 5.1** Example of long-term aerobic training protocol, by the authors

| Stage | Date of start | Warm-up | Active phases | Recovery phases | Cool down | Total tr. time | Stage duration |
|---|---|---|---|---|---|---|---|
| 1 | | 3 min | 4 × 4 min | 3 × 2 min | 3 min | 28 min | 3 wks |
| 2 | | 3 min | 5 × 4 min | 4 × 2 min | 3 min | 34 min | 3 wks |
| 3 | | 3 min | 6 × 4 min | 5 × 2 min | 3 min | 40 min | 3 wks |
| 4 | | 3 min | 3 × 6 min | 2 × 2 min | 3 min | 28 min | 4 wks |
| 5 | | 3 min | 4 × 6 min | 3 × 2 min | 3 min | 36 min | 4 wks |
| 6 | | 3 min | 5 × 6 min | 4 × 2 min | 3 min | 44 min | 4 wks |
| 7 | | 3 min | 3 × 8 min | 2 × 2 min | 3 min | 34 min | 5 wks |
| 8 | | 3 min | 4 × 8 min | 3 × 2 min | 3 min | 44 min | 5 wks |
| 9 | | 3 min | 5 × 8 min | 4 × 2 min | 3 min | 54 min | 5 wks |
| 10 | | 3 min | 3 × 10 min | 2 × 2 min | 3 min | 40 min | 6 wks |
| 11 | | 3 min | 4 × 10 min | 3 × 2 min | 3 min | 52 min | 6 wks |
| 12 | | 3 min | 2 × 15 min | 1 × 2 min | 3 min | 38 min | 8 wks |
| 13 | | 3 min | 3 × 15 min | 2 × 2 min | 3 min | 55 min | 8 wks |
| 14 | | 3 min | 2 × 20 min | 1 × 2 min | 3 min | 48 min | Continued |

## 5.2 Exercise Centres

The concept of exercise centers for cardiac patients or individuals with cardiovascular risk factors is built on regular physical activity's beneficial role in primary and secondary cardiovascular prevention. These benefits include:

- Weight management
- Blood pressure control
- Lipid management
- Blood glucose control
- Stress reduction
- Improvement of left ventricular systolic and diastolic performance

**Training heart rate (trHR)**

| Enter data | | | Result trHR | |
|---|---|---|---|---|
| restHR (bpm) | peakHR (bpm) | Intensity (%) | %HRR | %peakHR |
| 72 | 128 | 55 | 103 | 70 |

Please do not enter any data in the yellow field

%HRR - percent heart rate reserve (by Karvonen)

**VO2 and MET's calculations for cycle ergometer according to ACSM**

| Enter data | | Results | |
|---|---|---|---|
| WR (watt) | BM (kg) | VO2 | MET's |
| 100 | 70 | 22.7 | 6.5 |

Please do not enter any data in the yellow field

WR - work rate

BM - body mass

VO2 - oxygen uptake (ml/min*kg)

**MET to watt calculation by ACSM**

*If the exercise test was done on a tredmill and You want to train on a cycle ergometer:*

| Enter data | | Results | | |
|---|---|---|---|---|
| BM (kg) | MET;s | WR (watt) | WR-10% | WR-20% |
| 70 | 6.5 | 100 | 86 | 71 |

***Attention:** Physical efficiency performed on a bicycle ergometer is 10-20% lower than on the treadmill*

**Fig. 5.1** Training heart rate calculator and MET to Watt converter, by the authors

Grade (incline) correlations of speed and MET's on a treadmill

| Enter data | | Result |
|---|---|---|
| S (km/h) | MET's | G (%) |
| 4,0 | 7,0 | 12,0 |

S - optimal training walking speed

G - grade, incline of the treadmill

Speed correlations of grade (incline) and MET's on a treadmill

| Enter data | | Result |
|---|---|---|
| G (%) | MET's | S (km/h) |
| 12,0 | 7,0 | 4,0 |

G - grade, incline of the treadmill

S - optimal training walking speed

**Fig. 5.2** Digital solution for optimal adjustment of treadmill speed and grade to trained MET, by the authors

- Antiarrhythmic effects
- Ischemic preconditioning

Individualized exercise training programs should focus on primary and secondary cardiovascular prevention, addressing cardio-fitness and cardiac rehabilitation. Additionally, these programs should include specific training for cancer survivors to improve their physical and psychological well-being during treatment. Exercise programming for cancer patients enhances physical capacity and quality of life and can help alleviate symptoms. When designing exercise plans for cancer patients, it is essential to consider the type of cancer, the treatment stage, and the individual's

**a**

**Hospital / Department name**

**TRAINING INTENSITY CALCULATOR for CARDIAC REHABILITATION**

Name: **Tom Runner** Age (years): **57** Data: **2024-02-28**
Body mass (kg) : **81** Height (cm): **173** Body mass index BMI ($kg/m^2$): **27.1**
Nutrition class : **overweight**

**Exercise stress test (ExT):**
Exercise test modality: **treadmill** Exercise Protocol: **ind. ramp** Exercise test time (sec): **630**

| ExT variables | Rest | Peak exercise |
|---|---|---|
| Heart rate HR (bpm) | 78 | 149 ( 91 % $HR_{max}$) |
| Blood pressure BP (mmHg) | 130 / 85 | 160 / 85 |
| Double product DP (bpm*mmHg) | 10140 | 23840 |

Exercise capacity: **10.7** MET - **190** (watt) **113**% reference value acc. to Morris ( **9.5** MET)
Capacity class: **Above average**
Reason for termination: **muscle fatigue**
Rate of perceived Exertion RPE ($Borg_{6-20}$): **16**
Exercise ECG: **No arrythmias / Ectopy / ST segment changes**

Training intensity classes according to American College of Sports Medicine ACSM and European Society of Cardiology ESC with individual load parameters for the tested person.

| Training intensity class (ACSM 2022) | Individual load parameters for the tested person | | | | |
|---|---|---|---|---|---|
| | THR (bpm) | Bicycle TWR (watt) | Treadmill trMET | Treadmill setting S (km/h) / G (%) | RPE $Borg_{6-20}$ |
| Light 30-39% | 99 - 105 | 57- 75 | 3.9 - 4.8 | 3.0 / 5.8 to 3.4 / 7.3 | 9 - 11 |
| Moderate 40-59% | 106 - 120 | 76 - 113 | 4.9 - 6.7 | 3.5 / 7.3 to 4.3 / 9.8 | 12 - 13 |
| Vigorous 60-89% | 121 - 141 | 114 - 170 | 6.8 - 9.6 | 4.4 / 9.8 to 5.4 / 13.0 | 14 - 17 |
| Near maximal to maximal ≥90% | 142 - 149 | 171 - 190 | 9.7 -10.7 | 5.5 / 13.0 to 5.8 / 14.1 | ≥18 |

THR – training heart rate (as a % HR reserve); TWR - training work rate (as a % of the peak work rate); trMET – training metabolic equivalent of task (as a % of the MET reserve); S - speed. G - grade; RPE - rate of perceived exertion

| Training intensity class (ESC 2020) | Individual load parameters for the tested person | | | | |
|---|---|---|---|---|---|
| | THR (bpm) | Bicycle TWR (watt) | Treadmill trMET | Treadmill setting S (km/h) / G (%) | RPE $Borg_{6-20}$ |
| Light <40% | 96 - 105 | 47 - 75 | 2.7 – 4.2 | 2.2 / 3.3 to 3.1 / 6.4 | 10 - 11 |
| Moderate 40 - 69% | 106 - 127 | 76 - 132 | 4.3 - 7.4 | 3.2 / 6.4 to 4.6 / 10.6 | 12 - 13 |
| High intensity 70 - 85% | 128 - 138 | 133 - 161 | 7.5 - 9.1 | 4.7 / 10.6 to 5.3 / 12.4 | 14 - 16 |
| Very high intens exercise >85% | 139 - 149 | 162 - 190 | 9.2 - 10.7 | 5.4 / 12.4 to 5.8 / 14.1 | 17 - 19 |

THR – training heart rate (as a % HR reserve); TWR - training work rate (as a % of the peak work rate); trMET – training metabolic equivalent of task (as a % of the peakMET); S - speed. G - grade; RPE - rate of perceived exertion

**Fig. 5.3** **a** Individually tailored exercise prescription solution for long-term exercise training with variety of exercise training protocols including 'patient's data translated into different exercise intensity thresholds and **b** proposal of individually tailored different exercise training protocols, by the authors

**b**

## EXAMPLES OF INDIVIDUAL CONTINUOUS AND INTERVAL TRAINING

**Light Intensity Continuous Training LICT (intensity 30 to 39%)**

| | | | |
|---|---|---|---|
| - training heart rate range HR [bpm]: | **99** | to | **105** |
| - bike work rate range WR [wat]: | **57** | to | **75** |
| - treadmill load range [MET]: | **3.9** | to | **4.8** |
| - treadmill setting S [km/h]/G [%]: | **3.0 / 5.8** | to | **3.4 / 7.3** |

Warm up and cool down: 2-5 min; intensity 10-15 watt (cycle); 1.6 km/h. G=0 (treadmill)

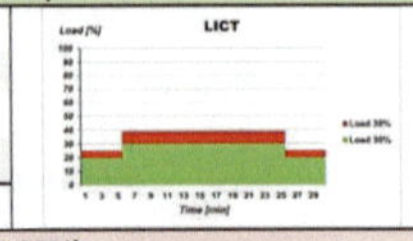


**Moderate Intensity Continuous Training MICT (intensity 40-69%)**

| | | | |
|---|---|---|---|
| - training heart rate range HR [bpm]: | **106** | to | **127** |
| - bike work rate range WR [wat]: | **76** | to | **132** |
| - treadmill load range [MET]: | **4.9** | to | **7.7** |
| - treadmill setting S [km/h]/G [%]: | **3.5/ 7.3** | to | **4.7/10.9** |

Warm up and cool down: 3-10 min; light intensity

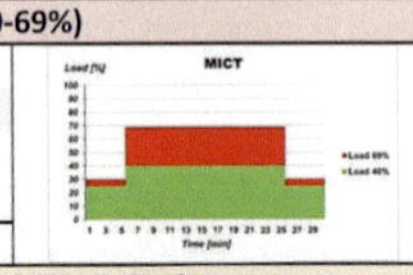


**Moderate Intensity Interval Training MIIT - work phase intensity 60-75% - 1-4 min**

| | | | |
|---|---|---|---|
| - training heart rate range HR [bpm]: | **121** | to | **131** |
| - bicycle work rate range WR [wat]: | **114** | to | **142** |
| - treadmill load range [MET]: | **6.8** | to | **8.3** |
| - treadmill setting S [km/h]/G [%]: | **4.4/ 9.8** | to | **5.0/11.5** |

Warm up and cool down: 3-10 min; light intensity

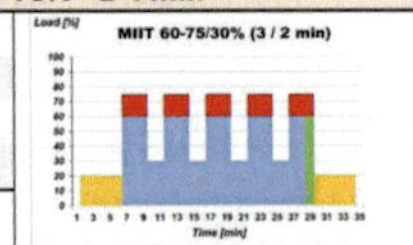


**High Intensity Interval Training HIIT - work phase intensity 85–100 (120%)**

| | | | |
|---|---|---|---|
| - training heart rate range HR [bpm]: | **138** | to | **149** |
| - bicycle work rate range WR [wat]: | **161** | to | **190 (228)** |
| - treadmill load range [MET]: | **9.2** | to | **10.7 (12.6)** |
| - treadmill setting S [km/h]/G [%]: | **5.3/12.5** | to | **5.8/14.1 ( 6.3/16.0)** |

**Work phase**: **s-HIIT** 85-100 (120%); <1 min (bicycle training 10. 15. 30 s);
**m-HIIT** 85-100 %. 1-3 min; **L-HIIT** 85-100%. 4 min . **Recovery phase**: passive **PR** or active **AR**.

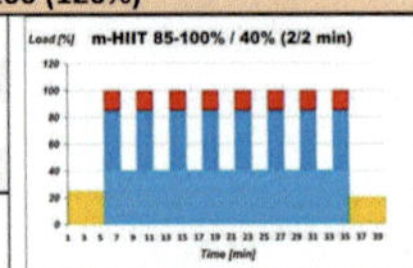


**MIIT. HIIT - Passive Recovery PR (with little or no load)**

| | | | |
|---|---|---|---|
| - training heart rate range HR [bpm]: | **78** | to | **96** |
| - bicycle work rate range WR [wat]: | **0** | to | **10 - 15** |
| - treadmill load range [MET]: | **1.7** | to | **2.1** |
| - treadmill setting S [km/h]/G [%]: | **1.5/0** | to | **2.0/0** |

Passive recovery time: 30 s – 3 min.

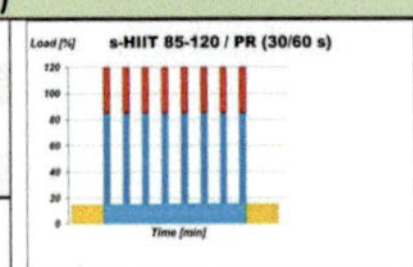


**MIIT. HIIT - Active Recovery AR 25-50%**

| | | | |
|---|---|---|---|
| - training heart rate range HR [bpm]: | **96** | to | **114** |
| - bicycle work rate range WR [wat]: | **47** | to | **95** |
| - treadmill load range [MET]: | **3.4** | to | **5.9** |
| - treadmill setting S [km/h]/G [%]: | **2.7/ 5.0** | to | **4.0/ 8.6** |

Active recovery time: 30 s – 3 min.

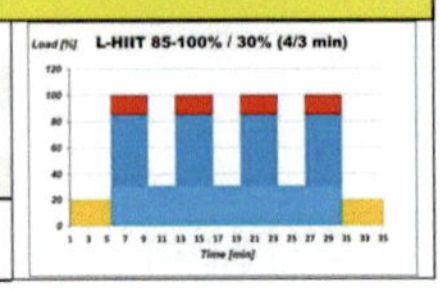


**Comment:**.............................................................................................................................................

..........................................................................................................................................................

..........................................................................................................................................................

........................................ ........................................

physiotherapist's signature doctor's signature

**Fig. 5.3** (continued)

overall clinical status. An exercise specialist should lead these programs in collaboration with an oncologist. It is crucial to prioritize the safety of these training programs, considering compromised immune systems and the risk of injury [5].

Exercise programming for individuals with cardiovascular diseases and/or risk factors should be closely monitored. This approach includes comprehensive functional capacity assessments using cardiopulmonary exercise testing and evaluations of cardiac performance through echocardiography, enhanced by speckle tracking echocardiography. This technique complements traditional echocardiography analysis and provides valuable insights into myocardial mechanics, enabling the early detection of subtle myocardial dysfunction and effective monitoring of treatment outcomes. Both modalities can be employed prior to major non-cardiac surgeries or after specific procedures to confirm improvements in cardiovascular health. The combination of cardiopulmonary exercise testing and echocardiography (CPET-ESE) is particularly beneficial as it can help differentiate between cardiac and pulmonary causes of exercise intolerance. It also reveals advanced stages of diastolic dysfunction, which is crucial for the early diagnosis of heart failure with preserved ejection fraction and provides insights into unexplained dyspnea [6]. Moreover, exercise centers should incorporate a relaxation component, such as a spa center, to support overall well-being. The development of such centers should be informed by research studies related to cardiac rehabilitation and should integrate artificial intelligence to create personalized exercise prescription programs. By adopting a holistic approach, these exercise centers aim to optimize cardiovascular health, enhance cardiovascular fitness, and improve overall quality of life. The concept of the exercise center is illustrated in Fig. 5.4.

An exercise intervention should be accompanied by dietary education and psychosocial support. Psychosocial support is essential for cardiac patients to address the psychological and social challenges associated with cardiac disease [7]. This support includes education and individual or group counseling sessions that help patients cope with anxiety, depression, and stress [8]. Dietary counseling for cardiac

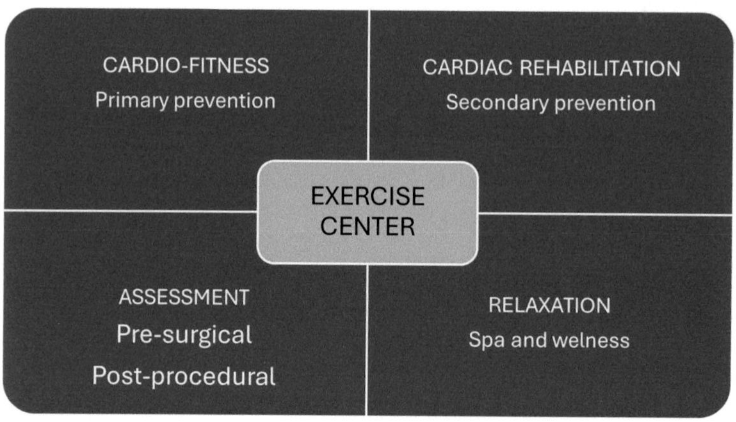


**Fig. 5.4** The concept of exercise centre, by the authors

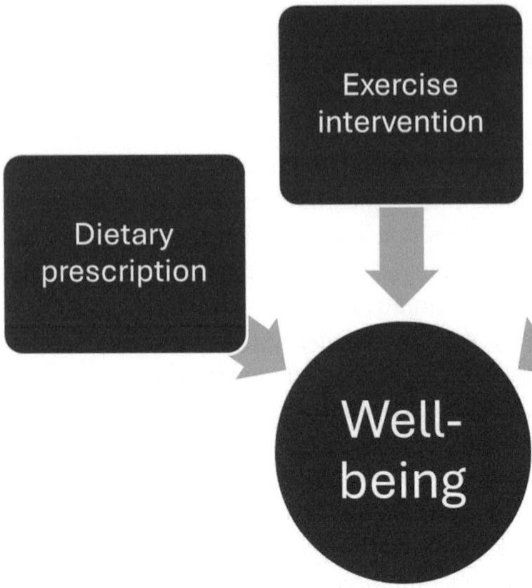


**Fig. 5.5** Holistic approach to exercise-based cardiovascular care, by the authors

patients focuses on promoting a healthy diet, providing meal planning, and long-term follow-up [9, 10]. This approach positively changes long-term diet and lifestyle adherence. The concept of a holistic approach to care is illustrated in Fig. 5.5.

**DataAvailability Competing Interests** The authors have no conflicts of interest to declare that are relevant to the content of this chapter.

## References

1. Piepoli MF, Hoes AW, Agewall S, Albus C, Brotons C, Catapano AL, Cooney MT, Corrà U, Cosyns B, Deaton C, Graham I, Hall MS, Hobbs FDR, Løchen ML, Löllgen H, Marques-Vidal P, Perk J, Prescott E, Redon J, Richter DJ, Sattar N, Smulders Y, Tiberi M, van der Worp HB, van Dis I, Verschuren WMM, Binno S. ESC scientific document group. 2016 European Guidelines on cardiovascular disease prevention in clinical practice: the sixth joint task force of the European society of cardiology and other societies on cardiovascular disease prevention in clinical practice (constituted by representatives of 10 societies and by invited experts) developed with the special contribution of the European association for cardiovascular prevention & rehabilitation (EACPR). Eur Heart J. 2016;37:2315–2381.https://doi.org/10.1093/eurheartj/ehw106.
2. Fletcher GF, Landolfo C, Niebauer J, Ozemek C, Arena R, Lavie CJ. Promoting physical activity and exercise: JACC health promotion series. J Am Coll Cardiol. 2018;72:1622–39. https://doi.org/10.1016/j.jacc.2018.08.2141.
3. Hansen D, Coninx K, Dendale P. The EAPC EXPERT tool. Eur Heart J. 20177;38):2318–2320. https://doi.org/10.1093/eurheartj/ehx396.
4. Hansen D, Dendale P, Coninx K, Vanhees L, Piepoli MF, Niebauer J, et al. The European association of preventive cardiology exercise prescription in everyday practice and rehabilitative

training (EXPERT) tool: a digital training and decision support system for optimized exercise prescription in cardiovascular disease. Concept, definitions and construction methodology. Eur J Prev Cardiol. 2017;24:1017–1031. https://doi.org/10.1177/2047487317702042.
5. Scott JM, Zabor EC, Schwitzer E, Koelwyn GJ, Adams SC, Nilsen TS, et al. Efficacy of exercise therapy on cardiorespiratory fitness in patients with cancer: a systematic review and meta-analysis. J Clin Oncol 2018.
6. Santoro C, Sorrentino R, Esposito R, Lembo M, Capone V, Rozza F, et al. Cardiopulmonary exercise testing and echocardiographic exam: an useful interaction. Cardiovasc Ultrasound. 2019;17:29. https://doi.org/10.1186/s12947-019-0180-0.
7. Lichtman JH, Froelicher ES, Blumenthal JA, Carney RM, Doering LV, Frasure-Smith N, Freedland KE, Jaffe AS, Leifheit-Limson EC, Sheps DS, Vaccarino V, Wulsin L. American heart association statistics committee of the council on epidemiology and prevention and the council on cardiovascular and stroke nursing. Depression as a risk factor for poor prognosis among patients with acute coronary syndrome: systematic review and recommendations: a scientific statement from the American heart association. Circulation. 2014;129:1350–69. https://doi.org/10.1161/CIR.0000000000000019.
8. Reavell J, Hopkinson M, Clarkesmith D, Lane DA. Effectiveness of cognitive behavioral therapy for depression and anxiety in patients with cardiovascular disease: a systematic review and meta-analysis. Psychosom Med. 2018;80:742–53. https://doi.org/10.1097/PSY.0000000000000626.
9. Trichopoulou A, Bamia C, Norat T, Overvad K, Schmidt EB, Tjønneland A, ey al. Modified Mediterranean diet and survival after myocardial infarction: the EPIC-Elderly study. Eur J Epidemiol. 2007;22:871–81. https://doi.org/10.1007/s10654-007-9190-6.
10. Lacroix S, Cantin J, Nigam A. Contemporary issues regarding nutrition in cardiovascular rehabilitation. Ann Phys Rehabil Med. 2017;60:36–42. https://doi.org/10.1016/j.rehab.2016.07.262.

MIX
Papier aus verantwortungsvollen Quellen
Paper from responsible sources
FSC® C105338

If you have any concerns about our products,
you can contact us on
**ProductSafety@springernature.com**

In case Publisher is established outside the EU,
the EU authorized representative is:
**Springer Nature Customer Service Center GmbH**
**Europaplatz 3, 69115 Heidelberg, Germany**

Printed by Libri Plureos GmbH
in Hamburg, Germany